NEXT GENERATION 9-1-1

NEXT GENERATION 9-1-1

EUGENE B. ZELIK
EDITOR

Nova Science Publishers, Inc.
New York

For permission to use material from this book please contact us:
Telephone 631-231-7269; Fax 631-231-8175
Web Site: http://www.novapublishers.com

LIBRARY OF CONGRESS CATALOGING-IN-PUBLICATION DATA

Zelik, Eugene B.
Next generation 9-1-1 / Eugene B. Zelik.
p. cm.
Includes index.
ISBN 978-1-60692-639-0 (hardcover)
1. Telephone--United States--Emergency reporting systems--Planning. 2. Telephone--Emergency reporting systems--Government policy--United States. 3. United States. Dept. of Transportation--Communication systems. I. Title. II. Title: Next generation 911.
TK6169.Z45 2009
384.6'4--dc22

2008055755

Published by Nova Science Publishers, Inc. ✢ New York

CONTENTS

PREFACE

Although the 9-1-1 system has been an unqualified success story for more than 30 years, changes in the public's use of technology, the saturation of the mobile market, and the spread of Voice over Internet Protocol (VoIP) telephony over broadband are contributing to greater expectations that the current system will need to address. Because text, data, images, and video are increasingly common in personal communications and are critical to future transportation safety, the 9-1-1 system will be expected to accommodate highly mobile, dynamic communications modes. The architecture of these communication nodes directly counters the fundamental structure of the current 9-1-1 system. To guide and foster a nationwide vision of a 9-1-1 system for the 21st century, the U.S. Department of Transportation (USDOT) is taking a lead role in the research and development needed to bring about a more capable Next Generation 9-1-1 (NG9-1-1) system that supports emergency call delivery and a response based system that maximizes impact across a diverse stakeholder community. There are, however, a large number of operational, economic, political, and institutional issues that must be addressed and reconciled if the NG9-1-1 system is to be implemented successfully across the nation.

Implementing NG9-1-1 will likely be a complicated process, requiring the effective, timely and willing cooperation of an array of stakeholders. To address these issues, the NG9-1-1 Initiative has begun researching and analyzing the strategic transition elements and options that could be followed to further the progress and implementation of NG9-1-1. This Preliminary Transition Plan will help frame how the government and industry view the deployment of NG9-1-1 and the key decisions that might need to be made by each level of government and the private sector to make implementation a reality.

Potential Deployment Approaches Transition to NG9-1-1 is expected to be evolutionary process, involving technological, economic, and institutional change. In some cases, the path to NG9-1-1 implementation will depend on the underlying infrastructure and state of the Public Safety Answering Point (PSAP) and 9-1-1 Authorities. In other cases, the transition to NG9-1-1 may depend on the ability of originating service networks to deliver NG9-1-1 calls via native internet protocol (IP)-based infrastructure to jurisdictions that are prepared to receive those calls.

In: Next Generation 9-1-1
Editor: Eugene B. Zelik
ISBN: 978-1-60692-639-0

Chapter 1

NEXT GENERATION 9-1-1 (NG9-1-1) SYSTEM INITIATIVE: NG9-1-1 PRELIMINARY TRANSITION PLAN

Intelligent Transportation Systems, U.S. Department of Transportation

OVERVIEW

Although the 9-1-1 system has been an unqualified success story for more than 30 years, changes in the public's use of technology, the saturation of the mobile market, and the spread of Voice over Internet Protocol (VoIP) telephony over broadband are contributing to greater expectations that the current system will need to address. Because text, data, images, and video are increasingly common in personal communications and are critical to future transportation safety, the 9-1-1 system will be expected to accommodate highly mobile, dynamic communications modes. The architecture of these communication nodes directly counters the fundamental structure of the current 9-1-1 system. To guide and foster a nationwide vision of a 9-1-1 system for the 21st century, the U.S. Department of Transportation (USDOT) is taking a lead role in the research and development needed to bring about a more capable Next Generation 9-1-1 (NG9-1-1) system that supports emergency call delivery and a response based system that maximizes impact across a diverse stakeholder community.

There are, however, a large number of operational, economic, political, and institutional issues that must be addressed and reconciled if the NG9-1-1 system is to be implemented successfully across the nation. Implementing NG9-1-1 will likely be a complicated process, requiring the effective, timely and willing cooperation of an array of stakeholders. To address these issues, the NG9-1-1 Initiative has begun researching and analyzing the strategic transition elements and options that could be followed to further the progress and implementation of NG9-1-1. This Preliminary Transition Plan will help frame how thegovernment and industry view the deployment of NG9-1-1 and the key decisions that might need to be made by each level of government and the private sector to make implementation a reality.

Potential Deployment Approaches

Transition to NG9-1-1 is expected to be evolutionary process, involving technological, economic, and institutional change. In some cases, the path to NG9-1-1 implementation will depend on the underlying infrastructure and state of the Public Safety Answering Point (PSAP) and 9-1-1 Authorities. In other cases, the transition to NG9-1-1 may depend on the ability of originating service networks to deliver NG9-1-1 calls via native internet protocol (IP)-based infrastructure to jurisdictions that are prepared to receive those calls. Regardless of the specific evolutionary steps, it is expected that NG9-1-1 system implementation within the public sector will likely follow one of two general frameworks for deployment:

Coordinated, Intergovernmental Approach: Planned and coordinated deployments of NG9-1-1 capabilities that are governed by statewide 9-1-1 Authorities, regional Authorities, or informal mechanisms that enable a cooperative deployment.

Independent, Unilateral Approach: Decentralized deployments of NG9-1-1 capabilities by local jurisdictions through independent initiatives.

The NG9-1-1 vision also implies some degree of national leadership for the deployment of next generation systems. While there is no single best national approach to coordinating NG9-1-1 implementation at a national level, potential stakeholders range from invested federal agencies and national associations.

Strategies and Options for NG9-1-1 Going forward

Although there are specific challenges and decisions for states, 9-1-1 Authorities, and PSAPs when choosing an implementation path, there are also national and local issues that need to be resolved irrespective of the transition approach. Consequently, the identification of broader strategic options for progressing toward NG9-1-1 has been defined in this document to help the emergency communications community and local, state, and federal policy makers address critical elements for success as they plan and implement NG9-1-1 nationwide. Within this document, the NG9-1-1 agenda for transition is defined by the following strategic elements: Funding, Operations, Standards and Technology, and Governance and Policy. Making progress in all aspects of these strategic elements is essential because the elements are largely interdependent. Therefore, to gain a true picture of a region's transition to NG9-1-1, progress along all of the elements of should be considered together.

As part of the transition planning process, the strategic elements described in this document will also signal which organizations will be responsible for responding to them and thus guide future options, decisions, and strategy. Within this Plan, each strategic element for the transition to NG9-1-1 is described in terms of what the strategic issues are, what has already been accomplished to address these issues, and what could still be done to address the issue. These elements are briefly described in Table E-1.

Table E-1. NG9-1-1 Strategic Elements

• Strategic Elements	• Description
Funding	• Most State and local Authorities obtain funding by imposing and collecting fees on telecommunications services such as landline, wireless, and IP-enabled voice services. Jurisdictions must be willing to commit the resources necessary to ensure the success of any next generation effort. For example, ongoing maintenance and support of the system must be planned for and incorporated into the budget. •
Operations	• Operations include PSAP operations, as well as broader standard operating procedures (SOPs), formal written guidelines and/or instructions for calls and incidents (e.g., PSAP goes down), as well as training, exercises, and operational optimization (e.g., business process reengineering, change management).

Standards and Technology	• Progress in each of the other elements is essential to proper planning, use, and implementation of NG9-1-1 technology, and should drive technology procurement. Technology is highly dependent on existing infrastructure within a region and/or what is available to states and PSAPs from service providers. Multiple technology solutions may be required to support unique situations.
Governance and Policy	• Governance and policy includes leadership and planning—both key to entities transitioning to NG9-1-1. For example, many regions face difficulties related to political issues and the relationships within and across jurisdictions and disciplines. Leadership can help to work through these challenging internal and jurisdictional conflicts as well as set the stage for a region's commitment to the NG9-1-1 effort. Success in this element is also based on public's awareness, use, and satisfaction with NG9-1-1 services.

NG9-1-1 Initiative Transition Planning: Nextsteps

Over the coming months, the NG9-1-1 Initiative will use the *NG9-1-1 Preliminary Transition Plan* to ensure the dialogue and cooperation established by the initiative is continued and extended in the future. All groups and individuals with a stake in NG9-1-1—industry, the general public, governments, communities, and nongovernmental organizations—have a right and a responsibility to have their voices heard in dialogue about the future of 9-1-1. The NG9-1-1 initiative will build on and expand its *Preliminary Transition Plan* throughout 2008, based on the feedback it receives after this document is published and released to the public. At the end of 2008, the NG9-1-1 Initiative will publish a final *NG9-1-1 Transition Plan* that serves as an agenda for action and a foundation for the 9-1-1 community in planning and deploying NG9-1-1.

Introduction — About this Report

NG9-1-1 Project Background

The Next Generation 9-1-1 Initiative (NG9-1-1) is a U.S. Department of Transportation (USDOT) research and development project that will help define the system architecture and develop a transition plan that considers responsibilities, costs, schedule, and benefits for deploying Internet Protocol (IP)-

based emergency technologies within the 9-1-1 network across the Nation.[1] As detailed in the *USDOT NG9-1-1 System Initiative: Concept of Operations (CONOPS)*, USDOT understands that access to emergency services provided by public safety answering points (PSAP) in today's world of evolving technology will ultimately occur within a broader array of interconnected networks comprehensively supporting emergency services—from public access to those services, to the facilitation of those services, to the delivery of the emergency information to dispatchers and first responders.[2]

The USDOT's NG9-1-1 Initiative, has two main areas of focus: Technical/Engineering and Institutional/ Transitional. The technical aspect of the NG9-1-1 Initiative mainly centers on documenting NG9-1-1 system requirements, developing a system architecture, and demonstrating a proof of concept system. The institutional aspect primarily focuses on developing an assessment of cost, value, and risk of the NG9-1-1 system and developing a transition plan to evaluate all non-technical factors (e.g., cost, stakeholders, impacts, benefits) that will affect the successful transition to NG9-1-1.

Purpose and Scope of the Preliminary Transition Plan

To facilitate the migration to the NG9-1-1 system, it is critical to understand and assess transition issues and identify potential options to resolve or address the issue. Without a clear understanding of the potential challenges and options to address the obstacles, the deployment of NG9-1-1 may extend over an inordinate length of time. The NG9-1-1 Preliminary Transition Plan summarizes key funding, standards and technology, operations, and governance and policy issues identified in the *NG9-1-1 Transition Issues Report* (Appendix A), that will affect the transition from today's 9-1-1 to a nationwide NG9-1-1 system. Furthermore, the Plan, at a high level, identifies potential strategic options from which each affected stakeholder type can select to mitigate or resolve the transition issue. This report serves as the initial step in developing an NG9-1-1 Transition Plan for the 9-1-1 community. Leveraging this report and extensive stakeholder engagements throughout the coming year,the USDOT will develop a final transition plan that further defines and discusses strategic options to assist the 9-1-1 community in transitioning to the NG9-1-1 system. In addition, the NG9-1-1 Transition Plan can inform public policy officials at all levels of government or on considerations for legislative and regulatory attention to ensure successful transition to NG9-1-1.

Methodology

The USDOT used a phased approach to ensure that the Preliminary Transition Plan captured a list of critical issues and associated potential strategic options related to the NG9-1-1 transition. In reviewing existing references, USDOT found a significant and growing body of work discussing and beginning to address NG9-1-1 transition issues, including testimony before legislative and regulatory bodies; standard development organization (SDO) documents; professional association and academic research, white papers, and issue papers; and lessons learned from statewide IP-based 9-1-1 system implementations (e.g., Indiana's IN9-1-1).

In addition to existing reference research, USDOT has or will continue to reach out to the 9-1-1 stakeholders, including users such as 9-1-1 Authorities, call takers, and the general public; associations; local, state, and federal government agencies; and vendors, including service providers and equipment providers, to verify the issues and options and provide feedback on the most effective method to present the final transition plan so that it is beneficial to the 9-1-1 community.

How to Use this Preliminary Transition Plan

The intended audience for this document includes 9-1-1 stakeholders who have interest in or are directly affected by the transition to NG9-1-1. This report seeks to address readers with varying levels of familiarity with NG9-1-1. For the 9-1-1 stakeholders who have interest in or are directly affected by the transition to NG9-1-1, this report provides a high-level description of issues that may arise and potential options that each stakeholder group can employ to mitigate the issue.

1. To learn more about the potential implementation paths that states and jurisdictions may follow to transition to NG9-1-1…

   Turn to the section How NG9-1-1 Could Be Implemented Across the Nation: Potential Deployment Approaches.

2. To learn more about the strategic options and transition elements that will enable these NG9-1-1 implementation paths…

   Turn to the section *Strategies and Options for NG9-1-1 Going Forward.* This section explains the key NG9-1-1 transition elements and how potential strategies and options have been analyzed within this preliminary report. The corresponding sections addressing Funding, Operations, Standards and Technology, and Governance and Policy describe, in more, detail what actions could be taken in the future to progress toward achieving NG9-1-1.

3. To find out how your organization or stakeholder group is affected and how it can help resolve key NG9-1-1 transition issues…

   Choose your role as a 9-1-1 stakeholder and the corresponding symbol from the section beginning on the next page. For example, if you are a 9-1-1 Authority, your symbol is shown at right. For each major transition element and strategy, you can look for your symbol within the *Strategies and Options* section or *Appendix D* to learn how your community may be able to *9-1-1 Authority Symbol* impact or contribute to the resolution of financial, operational, technology or governance issues.

Taken together, both the transition paths and strategies for addressing transition elements can help stakeholders better understand what options are available and who else may be involved in addressing issues resulting from the transition to NG9-1-1. Once all stakeholders understand what options are available and who else is affected, they can design or contribute to a tailored path that is appropriate for their jurisdiction.

BACKGROUND: FROM 9-1-1 AND E 9-1-1 TO NG 9-1-1

The Nation's 9-1-1 system architecture, built in the late 1960s, is based on a decades-old, analog, circuit-switched network technology. In the beginning, 9-1-1 call processing was basic, using three-digit dialing, circuit-based transmission to PSAPs, and neither Automatic Number Identification (ANI) nor Automatic Location Identification (ALI) functions were available. The calls were received and processed through wireline infrastructure to the PSAP. As new technologies were introduced, modifications were made to establish connection to this basic 9-1-1 infrastructure. "Enhanced" 9-1-1 (E9-1-1) enabled the functionality to route calls to appropriate PSAPs based on the location of the caller using selective routing equipment, as well as provide PSAPs with the number (ANI) and address information associated (ALI) with the caller.

In the mid-1990s, with the proliferation of wireless technologies, the 9-1-1 system faced new challenges—receiving wireless 9-1-1 calls and identifying the location of the caller. The Federal Communication Commission's (FCC) Order

94-102[3] proceeding established requirements and deadlines in two phases, requiring that all wireless carriers provide the caller's telephone number and location information to the PSAPs. In 2005, with the growth of Voice over Internet Protocol (VoIP) telephony, the FCC issued Order 05-116[4] requiring VoIP providers to offer 9-1-1 services to all subscribers. These new technologies required further modifications to the existing 9-1-1 infrastructure to enable the connection of wireless and VoIP 9-1-1 calls.

The growing consumer market penetration of both wireless and VoIP telephony, and the increasing use of advanced technologies they represent, has underscored the limitations of the current 9-1-1 infrastructure. The Nation's 9-1-1 system, based on decades-old technology, cannot handle video, text, images, and other data that are increasingly common in personal communications. The pace of change in technology will not slow. If government wants to ensure that the general public has access to 9-1-1 from multiple communications devices, it will need to ensure that the 9-1-1 infrastructure can accommodate new technologies. If left unchanged, the current 9-1-1 system will face a multitude of pressures as society and technology continues to advance. For example—

At least 8 million customers currently rely solely on wireless communications as their primary telephone service (having discontinued wireline service or chosen not to use it).[5]

Estimates are that 12–15 million households will be using a VoIP service as either a primary or secondary telephone line by the end of 2008.[6]

Current 9-1-1 System Deficiencies

For a member of the public making an emergency call from a traditional wireline telephone, the 9-1-1 system works as designed. However, the proportion of calls to 9-1-1 calls placed from wireless telephones is approaching one-half in many communities.[7] A small, but growing, number of telephone users have acquired VoIP service. Wireless and VoIP service users often do not have traditional wireline telephone service in their homes or offices. As a result, the 9-1-1 system is becoming functionally deficient for this growing segment of the public. In addition, the system as it exists cannot incorporate the multiple types of data transmitted by new communications devices, and therefore, cannot unlock the potential for better emergency service delivery that these data types represent.

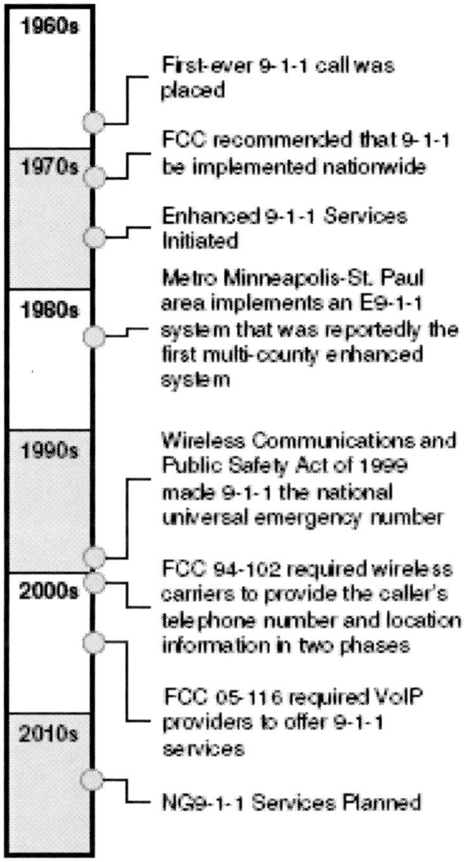

The communications and computing technology available to the public has advanced and continues to advance faster than the 9-1-1 system is able to change to meet the consumer's needs.[8] In particular, technology has allowed 9-1-1 callers and their communications devices to be mobile, thus creating a major challenge for the current 9-1-1 service delivery model, which relies on accurate callback and location data for every 9-1-1 call. It has now been more than a decade since the FCC required that all wireless carriers provide the caller's telephone number and location information to the PSAPs, yet not all PSAPs can receive this information from wireless calls.[9] With the advancement of technology, the emergency communication networks are continuing to become less efficient, less technologically advanced, and, as a result, less able to provide the public with 9-1-1 services on newer technologies and devices.

In addition, the deaf, hearing, and speech-impaired segment of the population has been historically underserved by 9-1-1. Current regulations require PSAPs to provide direct and equal access to their services to all citizens, regardless of disability.[10] As a result, PSAPs operators employ Telecommunication Device for the Deaf/Teletypewriter (TDD/TTY) equipment to communicate with the deaf and hearing-impaired, and assistive technologies to aid speech-impaired individuals. However, although the technology used by individuals with a disability has improved dramatically through the use of consumer devices, access to 9-1-1 has not significantly improved, and the population has become even more underserved by the Nation's PSAPs.

NG9-1-1: Where We Are Going…

USDOT and the 9-1-1 community believe that a technological evolution to NG9-1-1 is essential for the Nation's 9-1-1 networks to adapt to the public's increasing use of wireless communications and digital and IP-based devices to communicate. Many of these devices also transmit video, text, images, and other data. These technologies enable major advances in the ability of callers and public safety responders to send and receive useful information to, from, and beyond the emergency services internetwork.[11] Two examples that illustrate this point are emergency calls in American Sign Language (ASL) via video, and medically relevant telematics data transmitted from a car after a vehicle crash. With the trends in telecommunication and information technology (IT) convergence,[12] the current 9-1-1 systems should transition to a state-of-the-art infrastructure that will enable the transmission of voice, text, images, and other data from all types of communications devices to PSAPs, and on to emergency responder networks.

NG9-1-1 will be part of a comprehensive emergency communication system (or "system of systems") that will capitalize on advances in information and communications technologies, and will enable—

Quicker and more robust information delivered to both responders and the general public as the result of making a 9-1-1 call

Better and more useful forms of information (text, images, and video) from any networkedcommunications device

Transfer of 9-1-1 calls between geographically dispersed PSAPs (and from PSAPs to remote public safety dispatch centers), if necessary

Increased aggregation and sharing of data, resources, procedures, and standards to improve emergency response

Maximized use of available public capital and operating cost savings for emergencycommunications services

Promotion of increased coordination and partnerships within the emergency responsecommunity.

The NG9-1-1 system provides an enhanced 9-1-1 service to the public allowing callers to request emergency assistance by sending text, images, and video (in addition to voice) from several different kinds of access networks and communications devices. The USDOT NG9-1-1 CONOPS defines the NG9-1-1 system as—

> ". . . an evolutionary transition to enable the general public to make a 9-1-1 "call" from any wired, wireless, or IP-based device, and allow the emergency services community to take advantage of E9-1-1 call delivery and other functions through new internetworking technologies based on open standards."[13]

9-1-1 Authorities will need to deploy NG9-1-1 solutions that accommodate legacy components as depicted in figure 1.

Stakeholders that Will Be Key to the NG 9-1-1 Transition

Transition to the NG9-1-1 system will be affected by a large and varied group of stakeholders ranging from the general public and local emergency responder agencies to federal health, security, and emergency management agencies. Early and continued participation in NG9-1-1 planning by all 9-1-1 stakeholder groups is critical to successfully deploying the NG9-1-1 system. Although the boundaries of the USDOT NG9-1-1 Initiative are limited to the delivery and processing of the 9-1-1 call, this document addresses all stakeholders, including service providers, responder agencies, legislative and regulatory bodies, SDOs, and non-governmental organizations (e.g., professional and industry associations, and citizen and special interest advocacy organizations) that can directly or indirectly influence the transition to NG9-1-1. To better understand the fundamental roles and functions of each 9-1-1 stakeholder, Table 1 below provides a high-level description of each stakeholder type. In this context, "type of stakeholder" refers to an entire stakeholder category rather than individual entities within that category (e.g., responder agencies versus individual first responders).

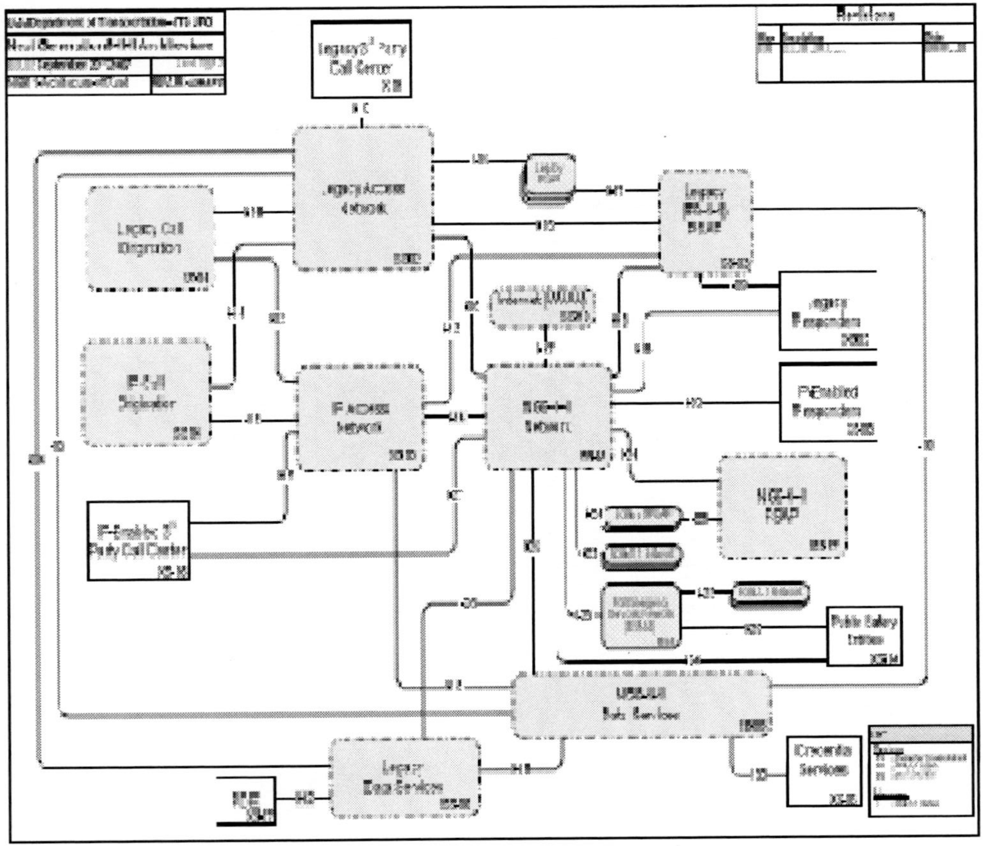

Figure 1. Composite NG9-1-1 High-Level System Architecture.[14]

Table 1. 9-1-1 Stakeholder Types

• Type of Stakeholder	• Symbol	• Description
• 9-1-1 Authorities and PSAP Administrators		• A 9-1-1 authority has jurisdiction over and/or supports a particular 9-1-1 system. The 9-1-1 authority could be a county/parish or city government, a special 9-1-1 or emergency communications district, a council of governments, an individual PSAP, a state agency, or other similar body. The 9-1-1 authority generally manages human resource requirements and activities, oversees service delivery, performs high-level database management and support functions, oversees funding and procurement of supporting infrastructure and services, establishes standard operating procedures (SOP) and operational policies, contracts with vendors and 9-1-1 service providers, and is responsible for security at the PSAPs. Along with the 9-1-1 Authority, the PSAP Administrator directs the overall operation of a PSAP and is responsible for the direct supervision, training, and administration of the PSAP's staff. The PSAP Administrator may be responsible for the maintenance of PSAP call-taking equipment and supporting peripherals. The PSAP Administrator may be responsible for the PSAP's budget and staff support.

Table 1. (Continued).

• Type of Stakeholder	• Symbol	• Description
• State Government (Legislative and Regulatory Bodies)		• Legislative bodies within each state draft and pass legislation for the welfare and benefit of the public. To ensure that the provisions of the legislation are implemented and enforced, regulatory bodies (e.g., Public Utility Commissions) may write draft regulations that are open to public comment but binding on the public once promulgated in final form. The regulatory body itself or a court having the appropriate jurisdiction may impose sanctions for regulatory infractions.
• Federal Government Agencies and Regulatory Bodies		• The U.S. Congress and federal government agencies are responsible for establishing national policies and funding, providing leadership, and promoting coordination and communications between agencies and organizations involved with 9-1-1 services. The most prominent government organizations involved in 9-1-1 include the USDOT Intelligent Transportation Systems (ITS) Joint Program Office (JPO), Department of Homeland Security (DHS), USDOT National Highway Traffic Safety Administration (NHTSA)–National Telecommunications and Information Administration (NTIA) National 9-1-1 Implementation and Coordination Office (ICO), Department of Justice (DOJ), FCC, and Department of Health and Human Services (HHS).

Table 1. (Continued).

• Type of Stakeholder	• Symbol	• Description
• Non-Governmental Organizations		• Non-governmental organizations (e.g., agencies and organizations that are considered part of the private sector) play a critical role in the implementation of the NG9-1-1 system because of their involvement in identifying the needs of their membership, thereby shaping the products and services used by consumers and PSAPs who use the NG9-1-1 system. Examples of non-governmental organizations include SDOs, Public Safety Communications Organizations and Associations, other professional and industry associations, citizen and special interest advocacy organizations, private emergency response and recovery organizations, and research and academia.
• Service and Equipment Providers & Third-Party Call Centers		• Service and equipment providers are responsible for developing, supplying, and sometimes maintaining the hardware, software, and other equipment used by PSAPs in transferring 9-1-1 calls; providing public access to 9-1-1 services, physical and network access, and telecommunications and IT infrastructure equipment; and helping to maintain the data and information necessary to route and deliver emergency calls. Types of service and equipment providers include 9-1-1 service providers, service access providers, service application providers, and equipment suppliers. Third-party call centers receive voice, video, text, images, and other data originating from the public via various services, such as automobile telematics, relay services, medical alert, poison control hotlines, alarm monitoring, and satellite telecommunications services.

Table 1. (Continued).

• Type of Stakeholder	• Symbol	• Description
Responder Agencies		• Responder agencies—dispatch entities, law enforcement, fire and rescue, EMS, supplemental responders, and beneficiary entities—have a legal or consensual obligation to respond to or obtain information from emergency calls to support comprehensive incident management.
General Public		• The general public relies heavily on the ability to access 9-1-1 service with full functionality through any communications device during an emergency. The public expects a timely response from the emergency responder agencies once a call is made to 9-1-1 and determines the success of the 9-1-1 service based on the response time. To date, the general public can make 9-1-1 calls through wireline telephone, wireless telephone, TDD/TTY over wireline, and VoIP, and indirectly through third-party call centers associated with telematics and relay services.

How NG9-1-1 Could be Implemented Across the Nation: Potential Deployment Approaches

What are the Potential Paths to NG9-1-1 Deployment?

As defined by the NG9-1-1 Initiative, the NG9-1-1 system is expected to be an interconnected system of local and regional emergency services networks ("system of systems").[15]However, the boundaries of emergency service networks may vary, depending on local requirements and organizational frameworks. Consequently, although the architecture and underlying requirements of NG9-1-1 will be relatively consistent, it is expected that there will be several implementation paths to NG9-1-1—all with merit.

Transition to NG9-1-1 is also expected to be evolutionary process, involving technological, economic, and institutional change. In some cases, the path to NG9-1-1 implementation will depend on the underlying infrastructure and state of the PSAP and 9-1-1 Authorities. In other cases, the transition to NG9-1-1 may depend on the ability of originating service networks to deliver NG9-1-1 calls via native IP-based infrastructure to jurisdictions that are prepared to receive those calls.

Regardless of the specific evolutionary steps, it is expected that NG9-1-1 system implementation within the public sector will likely follow one of two general frameworks for deployment:

Coordinated, Intergovernmental Approach: Planned and coordinated deployments of NG9-1-1 capabilities that are governed by statewide 9-1-1 Authorities, regional Authorities, or informal mechanisms that enable a cooperative deployment.

Independent, Unilateral Approach: Decentralized deployments of NG9-1-1 capabilities by local jurisdictions through piloting independent initiatives.

The following sections describe the key elements of these approaches, the assumptions under each approach, as well as an analysis of the transition issues that are specific to that implementation approach. It is expected that several transition issues will need to be resolved regardless of the specific deployment approach. Transition issues that are independent of a jurisdiction's implementation approach are addressed in the section entitled *Strategies and Options for NG9-1-1 Going Forward.*

The Coordinated, Intergovernmental Approach

What Is this Deployment Approach?

Under a coordinated, intergovernmental deployment approach, planned and coordinated deployments of NG9-1-1 capabilities are governed by statewide 9-1-1 authorities, regional authorities, or informal mechanisms that enable a cooperative deployment. Over time, many states have already developed intergovernmental or interjurisdictional mechanisms to coordinate, plan, and help guide the delivery of 9-1-1 services across jurisdictions. These mechanisms range from regional and metropolitan institutional frameworks to statewide programs vested in state government. The goals of these efforts have largely been designed to help address consistent and coordinated 9-1-1 service delivery throughout regions and states, along with the need to address telecommunication services that are becoming much more mobile and complicated. While most of these efforts are supported through enabling legislation, in some cases, local 9-1-1 Authorities have worked out other, less formal ways to address the need for coordination and joint planning.

Within a coordinated, intergovernmental approach, there are several ways an NG9-1-1 system could be implemented. While all involve intergovernmental coordination and planning to some degree, they vary in institutional arrangements, support, and authority based on a variety of factors related to the historical relationship of state and local government and the evolution of 9-1-1 services within the states involved. Generally, this approach can take one of three forms: 1) single statewide 9-1-1 authority; 2) regional authority; and/or 3) informal mechanisms that enable a cooperative NG9-1-1 deployment. These are illustrated in Figure 2 and are defined below.

Single Statewide 9-1-1 Authority: Where such authorities exist, they are usually created by statute, executive order, or, operate as a function of existing authority (like a state utility commission, for example). Authority generally covers both funding and implementation, although the latter ranges from statewide systems and outright authority over PSAP service migration and enhancement, to limited powers of encouragement, coordination, and guidance of same.

Regional Authorities Without a Statewide Authority: Some states do not provide for state-level 9-1-1 Authorities; rather they authorize local governments to join together in regional efforts to coordinate service delivery and share resources. That authorization may occur through specific enabling legislation (directed toward 9-1-1), or be a by-product of existing joint powers/inter-local cooperation legislation. In any case, such organizations have a legal basis of existence and may operate as political subdivisions of the state.

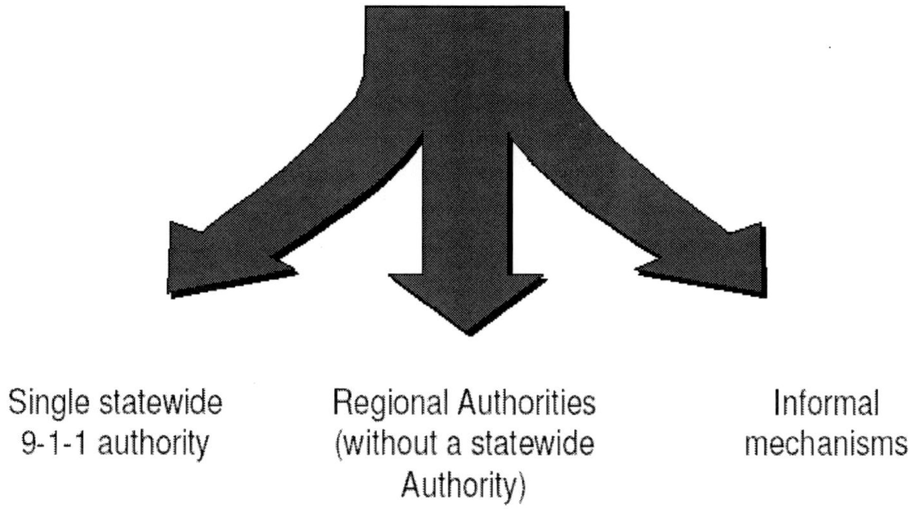

Figure 2. Types of Coordinated/Intergovernmental NG9-1-1 Deployments.

Informal Mechanism, Local Initiated Coordination: Generally, local initiatives coordinate 9-1-1 services and enhancement across traditional jurisdictional boundaries, where more formal mechanisms do not exist. These efforts vary from endeavors focused on metropolitan regions, to statewide initiatives. Normally they are more informal in nature (e.g., conducted through regional and state associations, and other cooperative endeavors), but can be quite effective depending on the organizations involved.

What Are Some of Assumptions and Key Dependencies of this Approach?

The common theme among all these forms of implementation is the desire to coordinate 9-1-1 service delivery between 9-1-1 Authorities. That may be as simple as working together in a structured, cooperative way, to employing a higher level coordinating body with the authority to ensure that coordination. The desire to coordinate does add an additional factor to NG9-1-1 migration. Authorities must balance local initiative and timing with that of the greater good. Consequently, decision authority, goals, and priorities are shared to some extent, and the process may take longer to implement. On the other hand, the process does provide the opportunity to share resources, deploy new and enhanced multi-jurisdictional service arrangements, and deal with emerging challenges in more effective ways (e.g., transferring 9-1-1 calls among jurisdictions, providing backup and mutual support, addressing larger scale emergency preparedness and homeland security needs, etc.).

What Would this Deployment Approach Look Like?

The State of Springfield has a statutory 9-1-1 function assigned to the state's General Services Commission (GSC). Under the statute, the GSC cannot force local governments to implement 9-1-1 services, but it can help fund, guide, and encourage such activity. In this state, there is a large metropolitan region with one central city and numerous suburban municipalities. The statute allows these governmental entities to join together in a multijurisdictional emergency communications district. While that district does not have the authority to directly operate PSAPs across the region, it can assess 9-1-1 fees and help fund such operation, along with providing connecting infrastructure, service, and support. Working together with the state, the district works out a plan for the region to collectively migrate to NG9-1-1 and to support state needs for public safety radio interoperability, emergency preparedness, and poison control.

In this scenario, the state provides statewide perspective and coordination, some additional funding to ensure that state needs are met, and state procured backbone infrastructure to facilitate the process. The district provides regional connecting infrastructure through an Emergency Services IP-based Network, facilitates expanded and focused service arrangements between local 9-1-1 Authorities, and supports regional database and related functions. The local 9-1-1 Authorities operate the PSAPs involved and provide emergency response services.

Which Transition Issues Are Different with this Approach than with Other Approaches?

Several issues will be different for the coordinated, intergovernmental approach because of the level of complexity or the opportunities that are available.

Governance and Policy: Underlying the coordinated, intergovernmental approach is the need for coordination among multiple 9-1-1 Authorities. With more entities involved, the roles and responsibilities and intergovernmental arrangements for the NG9-1-1 system will need to be defined across jurisdictional boundaries and between new partnerships. This is more complicated than the independent, unilateral approach because of the potential need to determine which entity is responsible for what, how to share different components of NG9-1-1, and how to address liability and confidentiality protection across state lines (assuming that liability and confidentiality protection may be different for each state)

Standards and Technology: With the connection between multiple PSAPs across jurisdictional boundaries, the control of system access and security may be more difficult to manage because of the number of entities involved. Furthermore, new processes and procedures may need to be defined to include all the jurisdictions and entities involved. In addition, the protocols and business rules for call routing and prioritization will be different and/or potentially more complicated with the number of PSAPs and jurisdictional coverage area involved and the number of potential options available to route calls (e.g., based on type of communications). However, it also provides the opportunity to consolidate PSAPs or promote the usage of "virtual PSAPs."[16]

Funding: Coordinating the implementation of 9-1-1 service delivery enables many 9-1-1 Authorities and PSAP Administrators to share various components of the NG9-1-1 system. This may provide opportunities for cost sharing and shared services among the 9-1-1 Authorities or other entities involved (e.g., emergency responders).

Independent, Unilateral Approach

What Is this Deployment Approach?

An Independent, Unilateral Deployment Approach to NG9-1-1 implementation entails a decentralized deployment of NG9-1-1 capabilities by local jurisdictions through piloting independent initiatives. The basis for this approach is that historically and traditionally, 9-1-1 services have been largely a local or municipal governmental responsibility. Indeed, the first assumption cited in the USDOT NG9-1-1 CONOPS notes:

> ". . . that the fundamental local institutional and operational frameworks for 9-1-1 services will remain in effect. That is, local government agencies will serve as answering points to receive, assess, and redirect 9-1-1 emergency calls from the general public to appropriate responders for help."[17]

In some states, that historical focus, coupled with a theme of strong local control of communications centers, has ensured that the responsibility for 9-1-1 service remains there as well., That may be difficult to change even with the need for larger scale geographical coordination and planning continuing to grow.

Consequently, this deployment framework will be found in a state with a 9-1-1 statutory environment that places 9-1-1 implementation responsibility solely on a local jurisdiction (e.g., at the township, city, or country/parish level). This type of situation can also exist where there is no state or regional governmental entity responsible for 9-1-1 services or where 9-1-1 statutes explicitly provide funding and implementation responsibility and authority to the local governments involved.

What Are Some of Assumptions and Key Dependencies of this Approach?

Under an independent, unilateral deployment approach, it is assumed that no other interjurisdictional coordination mechanism or initiative exists, and that efforts to migrate to NG9-1-1 will be limited to the separate initiatives by local jurisdictions. While the authorities may migrate to the NG9-1-1 system as described above, such improvements are limited to their jurisdiction and involve little if any service/infrastructure sharing or interconnection with neighboring entities (or, for that matter, with any other jurisdiction). Consequently, implementation in this scenario depends only on local initiative and resources and does not require time or effort to coordinate with a third part, nor the potential compromise that may require. Priorities can be set and migration managed as the jurisdiction best sees fit.

The general public would be able to make a 9-1-1 call from any kind of device, and call takers would be able to take advantage of other NG9-1-1 functionality at the PSAP, including expanded call treatment, access to additional information and data, and enhanced support of emergency response and incident management. Infrastructure and resource sharing is limited to the local public safety agency itself, as are the opportunities for innovative and expanded service arrangements with neighboring jurisdictions. While the authority may be able to implement new operational approaches like the so-called "virtual PSAP,"[18] arrangements involving other jurisdictions (e.g., coordinated incident management, support and backup,, emergency response, etc.) are necessarily limited by the willingness of the jurisdictions to work together.

What Would this Deployment Approach Look Like?

"Central City" in "Washington" County operates one PSAP and, under state statute, supports and funds it. Recently the City has made the decision to implement NG9-1-1. Working with a variety of infrastructure, equipment, and service providers, the City procured and deployed an IP-enabled NG9-1-1 system that offers many of the features described by nationally accepted next generation standards. They can accept 9-1-1 calls of all types (both in origination, and in format), process those calls in accordance with established business rules, and pass on enhanced data and information useful for emergency response and incident management.

However, based on long-standing precedent and history, the City does not coordinate much of what it does with the Sheriff's Office and the County. Inside the City, it is their business. Outside, it is the County's business. Nor does Central City work much with the "City of Farfield" in the neighboring county. Over the years, Central City has always been more progressive and reluctant to impair that progress by trying to do something in concert with a less progressive community like Farfield. Consequently, the City does not consider an option provided by the local community college to share in the cost of multifunctional IP-access network serving both counties and both cities, and deploys its own one-purpose, dedicated network. Nor does the City attempt to work out enhanced mutual aid and service arrangements with either Washington County or the City of Farfield. The City feels that such arrangements are generally more detrimental than beneficial, and, after all, it is able to receive and successfully serve the majority of calls it does receive.

Which Transition Issues Are Different with this Approach than with Other Approaches?

Several issues will be different for the independent, unilateral approach because of the level of complexity and the opportunities that are available.

Governance and Policy: With the independent, unilateral approach, it is expected that the 9-1-1 Authority responsible for a specific local jurisdictional area will remain the same, which means that governance and existing policies may either remain unchanged or only slightly change to accommodate NG9-1-1. In addition, it may be difficult to coordinate large-scale emergency response events because of the limited coordination between PSAPs across jurisdictional boundaries.

Standards and Technology: As 9-1-1 Authorities upgrade to NG9-1-1 without consideration for neighboring jurisdictions, it may be more complicated in the future to coordinate call treatment processes and response efforts that involve many jurisdictions. Call routing process and other SOPs may not exist or do not take into account neighboring jurisdictions to support the coordinated efforts.

Funding: As noted in the NG9-1-1 Transition Issues Report, funding may be limited because of the lack of priority of 9-1-1 upgrades and outdated funding models and allocation mechanisms. Consequently, implementing NG9-1-1 independently without the ability to share costs may make it difficult to identify sufficient funding to fully upgrade to NG9-1-1. Furthermore, the federal government is promoting public safety entities to coordinate upgrading efforts with neighboring jurisdictions to reduce cost through resource sharing, as well as to encourage interoperability.

Coordination and Implementation at a National Level

The vision for NG9-1-1 calls for an interconnected, nationwide implementation. To be sure, the USDOT NG9-1-1 Project CONOPS document, for example, speaks to ". . . a vision of a nationally interoperable emergency services internetwork." Consequently, this vision implies some degree of national leadership for the deployment of NG9-1-1—especially when compared to the implementation of E9-1-1 nationwide over the last decade. In recent legislation, Congress stated, "Enhanced 9-1-1 is a high national priority, and it requires Federal leadership, working in cooperation with state and local governments and with the numerous organizations dedicated to delivering emergency communications services.[19] "Without focus and leadership at a national level, NG9-1-1 could face challenges in realizing its goal of a national interconnected system.

There is no single best national approach to coordinating NG9-1-1 implementation at a national level. Potential stakeholders range from invested federal agencies like USDOT, NTIA and the FCC, to national associations like NENA, APCO and others. USDOT and NTIA jointly support the federal Implementation Coordination Office (ICO). Pursuant to existing legislation, the ICO defines its mission as ". . . [providing] leadership and coordination of comprehensive and technologically-enhanced [9-1-1] services." Specific responsibilities include improving coordination and communication between

Federal, State and local emergency communication systems, emergency personnel, public safety organizations, telecommunications carriers, and telecommunications equipment manufacturers and vendors. Also included are the development, collection and dissemination of information concerning practices, procedures and technology used in implementation of E9-1-1 services. Pending legislation would clarify the ICO's responsibility to include support for the 9-1-1 community's "migration to an IP-enabled emergency network."

At a higher level, the federal government also has a stakeholder interest in the connection between 9-1-1 and federal programs and services, including things like public safety radio interoperability, transportation safety and incident management. Beyond that, organizations like NENA, APCO and NASNA have a national association member interest in improving 9-1-1 services and migrating to next generation systems. The same is true of a variety of standards development organizations. All of this would benefit from some degree of national coordination and oversight.

Finally, there is the potential for a national organization to foster migration of 9-1-1 authorities and PSAPs using a national-level IP backbone and/or national-level NG9-1-1 services (e.g., similar to the implementation of National Law Enforcement Telecommunications System). However, this involvement may or may not occur. Initially, commercial service providers will likely provide the bulk of national or regional level IP-networks for the NG9-1-1 System; however, the implementation of these networks may take several years to fully resolve the regulatory issues involving security, database, and access rights.

Transition Issues Beyond the Deployment Approaches

The extent to which 9-1-1 Authorities or local jurisdiction move toward NG9-1-1 will be affected by underlying funding, operational structure, governance and policy, and the availability of appropriate standards and technology for IP access network infrastructure and next generation services and functions. As previously discussed, there are two potential deployment frameworks that 9-1-1 Authorities and other governing entities can employ to move from the current 9-1-1 System to NG9-1-1. Although the complexity of the issues may differ between the two frameworks, there are overarching issues that 9-1-1 Authorities and other governing entities will face before and during NG9-1-1 implementation regardless of which deployment framework is followed. High-level discussions of transition issues and associated strategic options have been identified in the following sections to assist 9-1-1 stakeholders in moving forward to implement NG9-1-1.

STRATEGIES AND OPTIONS FOR NG9-1-1 GOING FORWARD

The Key NG9-1-1 Transition Elements

Although there are specific challenges and decisions for states, 9-1-1 Authorities, and PSAPs when choosing an implementation path, there are also national and local issues that will be need to be resolved irrespective of the transition approach. Consequently, the identification of broader strategic options for progressing toward NG9-1-1 has been defined in this document to help the emergency communications community and local, state, and federal policy makers address critical elements for success as they plan and implement NG9-1-1 nationwide. The *NG9-1-1 Transition Issues Report* identified many of the key challenges that the Nation—at all levels of government—needs to overcome to ensure the achievement of the NG9-1-1 vision. These challenges influenced the design of potential strategies captured as high-level transition elements. These strategic elements represent the translation of the broad sense of direction stemming from the transition planning process into a practical set of options that can move the Nation to successfully implement NG9-1-1. They are the culmination of the transition planning process and describe the strategic agenda for the Nation as derived from the analyses of key transition issues.

Within this document, the NG9-1-1 agenda for transition is defined by the following strategic elements: Funding, Operations, Standards and Technology, and Governance and Policy. Making progress in all aspects of these strategic elements is essential because the elements are largely interdependent. Therefore, to gain a true picture of a region's transition to NG9-1-1, progress along all of the elements of should be considered together. For example, when a 9-1-1 Authority begins to procure new infrastructure, that jurisdiction should plan training and conduct testing to make the best use of that equipment. These elements are briefly described in Table 2.

Table 2. NG9-1-1 Strategic Elements

• Strategic Elements	• Description
Funding	• Most State and local Authorities obtain funding by imposing and collecting fees on telecommunications services such as landline, wireless, and IP-enabled voice services. Jurisdictions must be willing to commit the resources necessary to ensure the success of any next generation effort. For example, ongoing maintenance and support of the system must be planned for and incorporated into the budget.
Operations	• Operations include PSAP operations, as well as broader SOPs, formal written guidelines and/or instructions for calls and incidents (e.g., PSAP goes down), as well as training, exercises, and operational optimization (e.g., business process reengineering, change management).
Standards and Technology	• Progress in each of the other elements is essential to proper planning, use, and implementation of NG9-1-1 technology, and should drive technology procurement. Technology is highly dependent on existing infrastructure within a region and/or what is available to states and PSAPs from service providers. Multiple technology solutions may be required to support unique situations.
Governance and Policy	• Governance and policy includes leadership and planning—both key to entities transitioning to NG9-1-1. For example, many regions face difficulties related to political issues and the relationships within and across jurisdictions and disciplines. Leadership can help to work through these challenging internal and jurisdictional conflicts as well as set the stage for a region's commitment to the NG9-1-1 effort. Success in this element is also based on public's awareness, use, and satisfaction with NG9-1-1 services.

Understanding the NG9-1-1 Transition Elements and Options

As part of the transition planning process, the strategic elements described in this document will also signal which organizations will be responsible for responding to them and thus guide future options, decisions, and strategy. In the following sections of the document, each strategic element for the transition to NG9-1-1 will be described in terms of what the strategic issues are, what has already been accomplished to address these issues, and what could still be done to address the issue. As illustrated in Figure 3 below, the potential options to address NG9-1-1 issues will provide a potential indicator of which stakeholder has the power to take future action.

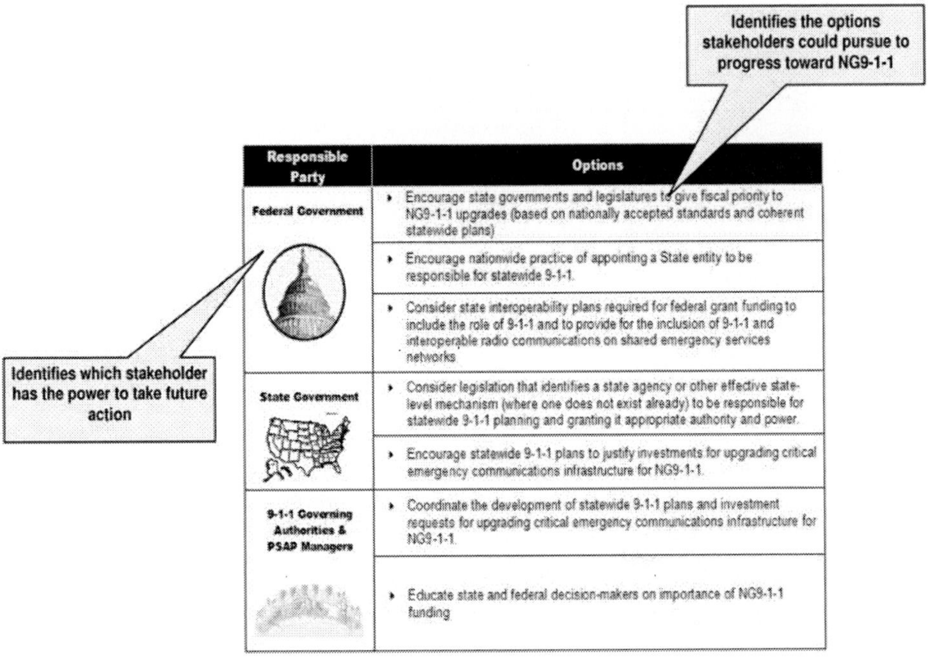

Figure 3. Example of NG9-1-1 Transition Options.

STRATEGIC OPTIONS FOR NG9-1-1 FUNDING

Background

Despite emerging requirements for the next generation of 9-1-1 services and capabilities, 9-1-1 Authorities and PSAPs throughout the Nation may struggle to finance new systems and capabilities while continuing to operate their current systems. Where existing revenue streams fall short, elected officials and government budget experts may face difficult choices when setting public investment priorities. These and other considerations make the funding environment for NG9-1-1 complex—recognizing and addressing these factors will be key to the successful migration to NG9-1-1:

NG9-1-1 as a Fiscal and Strategic Priority. Emergency communications networks and NG9-1-1 systems are costly. If dedicated and sustainable funding mechanisms are not available, these projects will face competition for funding from other worthy public projects. Many states, however, lack forward-looking strategic plans or roadmaps for planning and building out the foundations of NG9-1-1. Capital and strategic planning is critical if decision-makers are to be armed with the facts they need to justify high-priority emergency communications projects.

Adequacy of Funding Mechanisms for NG9-1-1. The concept of a 9-1-1 subscriber fee on telephone services remains virtually unchanged today from its inception. There is little consistency among and within states on 9-1-1 surcharge rates, which often differ based on the service type (e.g., wireless, wireline, or VoIP) or the jurisdiction for which the fee is being collected.

NG9-1-1 Cost Allocation and Cost-Reduction Strategies. In an NG9-1-1 environment, allocating costs will be a challenge, especially for costs related to shared networks, interfaces between systems, and costs for shared facilities and resources. Determining the appropriate allocation of costs will depend on the unique circumstances of a particular NG9-1-1 implementation.

Diversion of Dedicated 9-1-1 Funding Needed for NG9-1-1. Although many states and local jurisdictions assess a surcharge or fee for 9-1-1 services, the funds collected are not always used solely for 9-1-1 systems or PSAP maintenance and upgrades. There have been numerous, highly publicized instances of diversion of funds for non-9-1-1 uses such as balancing state budgets and funding state police. If states and local jurisdictions continue to divert 9-1-1 funds for other purposes, it will be more challenging and take longer to transition to NG9-1-1.

What Has Been Done?

Recently, some progress has been made to partially resolve elements of the NG9-1-1 funding issue:

Identification of Alternative Funding Models for NG9-1-1. The NationalEmergency Number Association (NENA) Next Generation Partner Programpublished a report. *Funding 9-1-1 into the Next Generation: An Overview ofNG9-1-1 Funding Model Options for Consideration* presents severalalternative funding models for consideration while outlining the benefits andpotential barriers to the proposed models. As noted in this initial study,funding models may not be exclusive because the best funding model for agiven deployment scenario may be a combination of several ideas.

Federal Legislation on 9-1-1 Fees. The Congress is considering legislation that makes clear that states and localities can impose fees on VoIP and future commercial mobile services, but that such funds must be expended in support of 9-1-1 and E9-1-1 and "enhancements to such services," which would include NG9-1-1.

Potential of Federal Grants and Loans for NG9-1-1. Increased interest in existing and proposed federal grant programs on the part of state and 9-1-1 Authorities is one sign that awareness regarding these alternatives is increasing. Grant programs administered by USDOT and the Department of Commerce are now being considered as a source of NG9-1-1 funding. Legislation under consideration in the Congress also seeks to broaden eligible use of federal PSAP grants to include "the migration to an IP-enabled emergency network" in addition to Phase II wireless E9-1-1. Legislation has also been introduced that would enable state or local governments to borrow from the Rural Utilities Service (RUS) to expand or improve 9-1-1 access and interoperable emergency communications.

State Awareness and Action. While changes in funding priorities have been slow in coming, other developments suggest that the issue of 9-1-1 fees and sustainable revenue are being recognized at the state and local levels. State legislative and executive bodies are more frequently considering the alteration of legislation that would address declining funding and changing call patterns.

What Could Be Done to Address NG9-1-1 Funding?

Additional options are available at all levels of government to address potential funding shortfalls that could affect nationwide migration to NG9-1-1. These options include potential strategies or paths that would—

Ensure NG9-1-1 upgrades are considered a fiscal priority for states and local jurisdictions and the Federal Government

Transform the current funding mechanisms to resolve the following issues:

Diminishing revenue base

Disparities with cost recovery

Funding allocation models for shared resources

Ensure 9-1-1 funds are preserved for 9-1-1 and emergency communication systems.

Ensure NG9-1-1 Upgrades Are Considered a Fiscal Priority for States and Local Jurisdictions and the Federal Government

Although the Federal Government considers emergency communications systems as critical infrastructure, it is a question whether the commitment exists at all levels of government to make migration to NG9-1-1 a fiscal priority. Gaining that commitment requires 9-1-1 Authorities to educate decision-makers and the public about the connection between NG9-1-1 planning and the public's continued access to lifesaving public safety services.

One of the key lessons learned from past E9-1-1 implementations is the role effective statewide coordination plays in focusing priorities for funding and support of the PSAPs. Applying this lesson to NG9-1-1 at the state level, governors can help lead statewide implementation of NG9-1-1 by designating a single point of contact for deployment, if one does not already exist. Such a coordinator would have authority to oversee implementation statewide resources. The Federal Government may also be able to help by encouraging centralized coordination and providing access to federal public safety grant programs. With access to public safety grant programs, state and local governments that do not believe they have the resources to dedicate to NG9-1-1 may have other options to obtain funding to continue planning and making progress as opposed to postponing or delaying NG9-1-1 deployment activities. Table 3 outlines some of the options government can take to help make NG9-1-1 a fiscal priority.

Transform Current 9-1-1 Funding Mechanisms to Resolve —

Diminishing revenue base

Disparities with cost recovery

Funding allocation models for shared resources

With the current revenue mechanisms and funding distribution, funding for NG9-1-1 implementation will be limited and further complicated during tight budget times. As more consumers abandon their wireline services and move to using wireless and VoIP technologies for their primary communications, overall revenues from the current funding model are decreasing. Consequently, changes are needed to sustain existing services while also advancing toward an IP-based NG9-1-1 system. Options that may be considered include creating technology-neutral revenue sources that reflect the complexity and diversity of the current telecommunications services and can more easily accommodate future services.

Table 3. Options to Improve the Fiscal Priority of NG9-1-1

• Responsible Party	• Options
• Federal Government 	• Encourage state governments and legislatures to give fiscal priority to NG9-1-1 upgrades (based on nationally accepted standards and coherent statewide plans).
	• Encourage states to designate a state agency or other effective state-level mechanism to be responsible for coordinating statewide 9-1-1, where such a mechanism does not already exist.
	• Consider requiring states to include 9-1-1 in their radio interoperability plans and to provide for shared emergency services internetworks.
• State Government 	• Consider legislation that identifies a state agency or other effective statelevel mechanism (where one does not exist already) to be responsible for statewide 9-1-1 planning and granting it appropriate authority and power.
	• Encourage statewide 9-1-1 plans to justify investments for upgrading critical emergency communications infrastructure for NG9-1-1.
• 9-1-1 Authorities 	• Coordinate the development of statewide 9-1-1 plans and investment requests for upgrading critical emergency communications infrastructure for NG9-1-1.
	• Educate state and federal decision-makers on the importance of NG9-1-1 funding.

Cost recovery and cost allocation models for NG9-1-1 will also vary by implementation approach. NG9-1-1 involves moving away from a closed and dedicated networks to a system with shared networks, databases, and applications in which costs related to the NG9-1-1 system are shared among all participants, some of which may be private sector entities or non-public safety governmental agencies. State governments and regulators can encourage shared services while also examining funding and cost recovery mechanisms and options. Table 4 outlines some of the options government can take to improve NG9-1-1 funding levels and mechanisms.

Table 4. Options to Improve NG9-1-1 Funding Levels and Funding Mechanisms

• Responsible Party	• Options
• Federal Government 	• Help States by providing grant subsidies for the capital costs of NG9-1-1 planning, design, procurement, and implementation.
	• Consider legislation that allows use of 9-1-1 fees to pay for NG9-1-1 portion and use of an underlying IP-based emergency service internetwork.
• State Government 	• Consider enacting legislation that imposes the 9-1-1 surcharge in a technologically neutral manner to accommodate current and future devices and services (e.g., text messaging, prepaid wireless).
	• Identify opportunities to offset the cost of NG9-1-1 by sharing infrastructure, resources, and services with, or simply interconnecting with, other public safety, non-public safety government, or private sector entities.
	• Consider enacting legislation that requires leveraging economies of scale to ensure efficient use of 9-1-1 revenues, and conducting annual audits on the use of the 9-1-1 funds.
	• Leverage alternative funding models (e.g., surcharge on access infrastructure provider [AIP]; user (incident) fee; Universal Statewide Communications Surcharge; General Fund tax revenue) that best suit a state's needs.
	• Consider legislation that allows 9-1-1 fees to be used to pay for the state's NG9-1-1 portion and use of IP-based emergency service internetwork.
	• Revisit how cost recovery is allocated to ensure fairness across all technologies and services, and determine whether cost recovery can or should be provided.
• Responsible Party	• Options
• 9-1-1 Authorities and PSAP Administrators 	•

Ensure 9-1-1 Funds Are Preserved for 9-1-1 and Emergency Communication Systems

The diversion of funding for 9-1-1 implementation is an ongoing issue that is not solely related to NG9-1-1 implementation. Regardless, if funds dedicated for 9-1-1 continue to be used for unrelated purposes then migration to NG9-1-1 will likely be affected. States and local governments that collect funds for 9-1-1 should use those funds solely for the purposes for which they were collected. In addition, if surplus funds should accumulate, these funds could be reserved for implementation of NG9-1-1 and 9-1-1's portion of an emergency service internetwork.

At the federal level, the Congress could consider providing an incentive by withholding federal 9-1-1 grant monies from states and political subdivisions that misappropriate, misallocate, and divert 9-1-1 monies from their intended purpose. A related option would be to require monitoring funds collection and usage and publishing information regarding the diversion of 9-1-1 funds from intended purposes. Table 5 outlines some of the options government can take to eliminate diversion of 9-1-1 dedicated funds.

Table 5. Options to Preserve 9-1-1 Dedicated Funds for Emergency Communications

• Responsible Party	• Options
	• Encourage state and local 9-1-1 Authorities to use 9-1-1 funds, surcharges, and fees for costs solely attributable to 9-1-1 operations, services, and equipment (or their statutory purposes).
	• Clearly define legislatively what constitutes 9-1-1 related communications services and the proper use of 9-1-1 fee revenue.
	• Consider tying eligibility for 9-1-1 grant funding to states that do not practice diversion of 9-1-1 funds. Consider mandating that 9-1-1 funds be used exclusively for the provision of 9-1-1 emergency communications services (e.g., 9-1-1, E9-1-1, NG9-1-1, and related or future upgrades).

Table 5. (Continued)

• Responsible Party	• Options
	• Require 9-1-1 funds to be used exclusively for the provision of 9-1-1 emergency communications services (e.g., 9-1-1, E9-1-1, NG9-1-1, and related or future upgrades).
	• Identify the appropriate uses of 9-1-1 funds and then monitor 9-1-1 Authorities to ensure 9-1-1 funds are used for costs solely attributable to 9-1-1.
	• Strengthen checks and balances to ensure funds are used only for the provision of 9-1-1 services, including shared services (e.g., 9-1-1 portion ofemergency services internetwork).

Responsible Why Does It Matter?

Funding for NG9-1-1 and emergency communications systems must be a national priority. Approximately half of the Nation's public safety agencies plan to replace their existing radio systems within the next 5 years. Along with public safety radio systems, 9-1-1 systems will also need to be upgraded as the public continues to adopt, as their primary means of communications, newer IP-based technologies that cannot access 9-1-1. Furthermore, to promote information sharing among emergency responders, 9-1-1 systems need to be on a technology platform similar to the systems used by public safety agencies. If funding impediments are not addressed, call takers and public safety agencies will become ill-equipped to handle current and future public safety communication requirements. Sufficient and sustained funding assures effective emergency communications for citizens in need and the public safety officials responding to those needs.

For Additional Information

As the NG9-1-1 system begins to become reality, additional information regarding funding can be found at—

9-1-1 Industry Alliance: 2008 Study on the Health of the US9-1-1 System

NENA: Funding 9-1-1 Into the Next Generation: An Overview of NG9-1-1 Funding Model Options for Consideration

STRATEGIC OPTIONS FOR NG9-1-1 OPERATIONS

Background

As implementation of the NG9-1-1 architecture continues, PSAPs will have to determine how to modify their existing practices and procedures and resources to implement and accommodate NG9-1-1 services. Several operational issues facing Authorities and PSAPs must be addressed, including—

New Expectations and Responsibilities of Call Takers. As a result of the introduction of NG9-1-1 technologies, PSAP operations will face new challenges associated with the increase in real-time multimedia information—text, still images, or video in addition to voice—passing between a person needing assistance and the call taker. The increased quantity of multimedia data will enhance and expand existing call-taking functions. Receipt of calls from IP-based communication devices will test call takers' ability to maintain their quick and accurate response to emergency situations.

Broader Operational Responsibilities for 9-1-1 Authorities. In addition to its effects on PSAP call takers, the NG9-1-1 environment will influence how 9-1-1 Authorities support PSAP operations and coordinate with other nearby 9-1-1 Authorities. IP-based communications create the potential for more seamless resource-and workload-sharing among formerly distinct PSAPs and 9-1-1 Authorities. 9-1-1 Authorities, especially those overseeing multiple PSAPs, will confront the challenge of managing a wider set of shared resources than is typical in the current system.

New Responsibilities and Challenges for PSAP Administrators.[20] At the local level, responsibilities of PSAP Administrators will likely expand, particularly with regard to configuring the NG9-1-1 system for their respective

PSAPs. Although PSAP Administrators may receive support and guidance from governing 9-1-1 Authorities, at the local level, these individuals will be on the front line for ensuring their specific facilities, staff, and resources meet the requirements and expectations necessary for implementing NG9-1-1.

Training and Certification Needs Will Expand Sharply. Although training programs for new call takers vary in duration, content, and format from PSAP to PSAP, the need for new training methodologies and curricula will become evident as NG9-1-1 technology is introduced. Continuing education (retraining in some cases) for experienced staff will be critical to the success of any NG9-1-1 implementation. Besides operational staff, technical and support staff will need the skills to configure, maintain, and troubleshoot advanced networks, systems, and components.

What Has Been Done?

There has been recent progress in addressing NG9-1-1 operational issues:

Education and Training for Call Takers. Future training needs and the framework for the curriculum needed for NG 9-1-1 training have been key topics of discussion throughout the PSAP community and public safety associations. As a starting point, existing education and training offerings and services are being reviewed by public safety communications organizations and associations to determine potential gaps and recommendations for training in the future.

Increased Awareness of NG9-1-1 Technology and Time Frames. NG9-1-1 is still considered a “new topic” by many in public safety communications. Associations and organizations have been delivering orientations and informational sessions at trade shows, conferences, and other meetings in an effort to educate communities and public safety personnel about what NG9-1-1 is, the features it can provide, and reference information to assist the locales in planning for NG9-1-1 implementation.

Standard Operating Procedures. 9-1-1 Authorities, trade associations, and other stakeholder groups are examining such issues as establishing requirements for development of remote PSAP functions, SOPs for handling 9-1-1 calls that are delivered with pictures and video, and SOPs for language services in a pre-next generation and next generation environment. NENA’s Accessibility and Standard Operating Procedures Committees’[21] activities are developing requirements and protocols for delivery and handling of text messaging.

NG9-1-1 Operations and Human Resources Requirements. 9-1-1 Authorities and PSAP Administrators are beginning to examine NG9-1-1 hiring, staffing, and scheduling issues. Work has also started in NG9-1-1 contingency planning, including backup, overflow, and redundancy issues. This contingency planning, along with development of requirements for "virtual PSAPs," is intended to assist PSAPs in building a comprehensive plan to test all aspects of their critical NG9-1-1 components through drills and exercises.

What Could Be Done to Address NG9-1-1 Operational Issues?

Additional options are available for stakeholders to address potential challenges that could affect PSAP operations. These options include potential strategies or paths that would—

Prepare and train call takers to handle increased quantity and quality of information available with a NG9-1-1 call

Prepare 9-1-1 Authorities to handle NG9-1-1 system administration, including configuration management, database management, quality assurance, and SOPs

Prepare 9-1-1 Authorities and PSAP Administrators to handle contingency planning and use of "virtual PSAPs."

Prepare and Train Call Takers to Handle Increased Quantity and Quality of Information Available with a NG9-1-1 Call

Although it is anticipated that the NG9-1-1 system will include human machine interface (HMI)[22] solutions to assist call takers in answering and processing call data, call takers may still need to analyze and make decisions regarding what information should be transferred to the dispatchers and responder agencies based on their training and experience. Consequently, in the NG9-1-1 environment, call takers will need to be trained to address call-handling activities such as answering multiple text messages from different callers during the same time period, and understanding text message forms such as the use of abbreviations, symbols, and short word sets. To date, many TTY/TDD 9-1-1 calls are not differentiated between voice 9-1-1 calls; however, with the NG9-1-1 System incoming calls can be divided between voice and TTY/TDD calls so that an assigned set of PSAPs or call takers within a PSAP would handle text messaging, while others handle voice calls.

To address these needs, associations like NENA and Association of Public-Safety Communications Officials (APCO) can work together to develop the model curriculum, protocols, and SOPs. These efforts can be supported through state and 9-1-1 authority training guidelines and approved training programs. PSAP Administrators will ultimately be responsible for implementing training and ensuring compliance, so their participation and feedback in the development of such material and recommendations will be key in refining and improving training and performance of NG9-1-1 call takers.

Table 6. Options to Address Call-Taker Training and Staffing for NG9-1-1

• Responsible Party	• Options
• Public Safety Communications Organizations and Associations 	• Develop guidelines for personnel skills/qualifications and effective training programs, including model training requirements for processing NG9-1-1 call (e.g., call handling, call treatment, and records management protocols).
	• Develop models for sharing data and managing information among PSAPs, public safety responders, and other authorized stakeholders.
	• Develop standardized common terminology that can be used by call takers similar to ongoing national standardization efforts involving various responder entities and others.
	• Consider conducting an annual survey to obtain input from call takers and their immediate supervisors on the effectiveness of training, SOPs, and system operations to ensure the quality of service is maintained.
• Federal Government 	• Consider conducting a nationwide annual survey to obtain input from call takers and their immediate supervisors on the effectiveness of training, SOPs, and system operations to ensure the quality of service for handling 9-1-1 calls is maintained.
	• Establish funding methods or a grant program to provide necessary financial support so users can be properly trained and outfitted with the necessary training materials.
	• Develop a public awareness program promoting public safety communications as a rewarding career opportunity in an effort to improve PSAP staffing levels.

Table 6. (Continued)

• StateGovernment 	• Where applicable, update or develop state-level standard training requirements for call takers (e.g., 40 hours of training to maintain certification).
• 9-1-1 Authorities and PSAP Administrators 	• Adopt training standards for processing NG9-1-1 call types, plans, implement training programs, and establish personnel qualifications.
	• Consider conducting a survey within the jurisdiction to obtain input from call takers on the effectiveness of training, SOPs, and system operations to ensure the quality of service for handling 9-1-1 calls is maintained.
	• Participate in and provide feedback on any state-level call-taker training.
	• Ensure regular training and exercises are completed at the PSAP.

Prepare 9-1-1 Authorities to handle NG9-1-1 System Administration, Including Configuration Management, Database Management, Quality Assurance, and SOPs

Much of the traditional 9-1-1 infrastructure is part of the existing telecommunications network, and its maintenance, upkeep, and oversight is typically handled by the dominant wireline telephone service provider in the region. Within NG9-1-1, the responsibility for the infrastructure, including its maintenance, upkeep, and oversight will likely fall under the oversight of a 9-1-1 Authority at the local, regional, or state level. Consequently, as NG9-1-1 is deployed, 9-1-1 Authorities will need to adjust and adapt to a broader set of responsibilities for far more complex technology systems.

To ensure 9-1-1 Authorities are able to support and manage networks and system resources that are shared across PSAP service areas, it will be important for entities like public safety organizations and the Federal Government to consider developing necessary tools and guidance. In some cases, these tools may be best practices for system configuration and administration. 9-1-1 Authorities, in turn, will need to establish the appropriate mechanisms and structure to accomplish the database and network functions necessary to support the NG9-1-1 enterprise.

Table 7. Options to Improve 9-1-1 Authorities' Ability to Manage NG9-1-1

• ResponsibleParty	• Options
• Public Safety Communications Organizations and Associations 	• Develop a generic Memorandum of Understanding (MOU) and other related templates so that 9-1-1 Authorities can work together (across state and jurisdictional limits) to determine processes for call overflows, backup conditions, and other automatic routing conditions.
	• Develop and gather implementation, operations, and maintenance best practices and standards, and share them among 9-1-1 Authorities.
	• Develop SOPs, protocols, and definitions for system configuration and management of different call types (e.g., text message).
• Federal Government 	• Develop implementation, and operations and maintenance best practices, standards, and lessons learned and share them among 9-1-1 Authorities.
	• Develop model strategic plans to help 9-1-1 Authorities at all levels manage NG9-1-1 migration (e.g., interagency coordination, training, security).
• 9-1-1 Authorities and PSAP Administrators 	• Develop and execute MOUs so that 9-1-1 Authorities can work together (across state and jurisdictional limits) to determine processes for call congestion, load sharing, backup conditions, and other automatic routing conditions.
	• Where needed, redefine roles and responsibilities for NG9-1-1 system administration.
	• Test the NG9-1-1 system to ensure that the level of quality of service is maintained based on terms and conditions agreed upon with the service providers.
	• Test the NG9-1-1 client application (e.g., HMI) to ensure that the level of quality of service is maintained based on terms and conditions agreed upon with the service providers.
	• Assess the amount of resources, both physical and human, necessary to provide NG9-1-1 coverage/service to an area based on population, number of calls, and other factors.
• Responders Agencies 	• Determine what data have the greatest utility in enhancing response and what processes will be used to deliver the data to ensure the right information gets to the right person at the right time.

Prepare 9-1-1 Authorities and PSAP Administrators to Handle Contingency Planning and Use of "Virtual PSAPs"

NG9-1-1 includes virtual PSAP capabilities, permitting flexible management for day-to-day operations and for disasters and major events. There are operational issues involved with this new concept and increased responsibilities for 9-1-1 Authorities and PSAP Administrators. For example, the concept of a virtual PSAP is one in which NG9-1-1 equipment serves multiple PSAPs with each PSAP having its own privileges and with a global administrator. These multiple PSAPs may be part of an intrastate or even interstate region. Consequently, in an NG9-1-1 environment, where physical geographic location is no longer a limiting factor, virtual PSAPs and networks of PSAPs provide additional potential resiliency that can be leveraged with contingency planning. To address these issues, organizations like NENA could provide valuable insight and guidance to 9-1-1 Authorities and PSAPs.

Table 8. Options to Address Contingency Planning and Virtual PSAPs

• Responsible Party	• Options
• Public Safety Communications Organizations and Associations 	• Develop requirements for virtual PSAPs.
	• Develop requirements and options for NG9-1-1 contingency planning (e.g., including backup, overflow, redundancy issues).
• Federal Government 	• Develop PSAP operations best practices and lessons learned and share them among 9-1-1 Authorities and PSAPs.
• 9-1-1 Authorities and PSAP Administrators 	• Develop requirements for virtual PSAPs within their jurisdiction.
	• Develop requirements and options for NG9-1-1 contingency planning (e.g., including backup, overflow, redundancy issues) for their jurisdiction.
	• Participate in operational and technical working groups to help define model requirements for virtual PSAPs and contingency requirements.
	• Hold regular combined training exercises to test the contingency plans and work to develop continuity plans of the desired level of detail.

Why Does It Matter?

NG9-1-1 will involve complex systems that may exceed existing system capacity, resources, and training. If such systems and their associated networks are not appropriately maintained and administered, key operational functions will be at risk. In extreme situations, this could inhibit the actual delivery of 9-1-1 calls. Resolution of NG9-1-1 operational issues require strategic decision-making regarding emergency service procedures, network management, database management, infrastructure, human capital, and facilities at all levels of government. These decisions will streamline procedures that improve the availability and accuracy of information, overall production and service, and effective emergency response and incident management.

For Additional Information

Although some work has been done within the community to establish strategic operational guidelines to transition to NG9-1-1, much work remains. There needs to be a coordinate effort to develop, share, implement, and refine these plans to maximize the PSAP's effectiveness. 9-1-1 Authorities should start or continue to heavily participate in training and educational opportunities, grant writing, public relations and awareness activities, and forging relationships with legislators and regulators at all levels of government (local, state, and federal). Knowledge and awareness of available opportunities is key, and that information is and will continue to be shared at meetings and conferences organized by a variety of organizations throughout the public safety community. Additional information regarding operations can be found at—

NENA: NENA IP-Capable PSAP Minimum Operational Requirements Standard

NENA: 2008 ODC Presentations, Documents and Notes

STRATEGIC OPTIONS FOR NG9-1-1 STANDARDS AND TECHNOLOGY

Background

As technology continues to evolve to support IP-based services for communications, such as wireless, digital landline, and VoIP advanced communications devices, the current 9-1-1 system is limited to receiving only the established set of ANI/ALI information, which is typically only 512 characters in length. Additional information, which is sometimes available today, is not readily accessible or provided to the call takers because of constraints in the legacy equipment. Standards for NG9-1-1 related technology and communications protocols are still in the process of being developed and finalized. The technological advancements and evolution of standards are independent of 9-1-1 and will drive the need for change in the 9-1-1 system. The ability for the public emergency communications infrastructure to keep pace with the continuing evolution of technology and standards is predicated on the ability to recognize and address key standards and technology issues to migrate to NG9-1-1:

NG9-1-1 Open Standards Development and Acceptance. The underlying concept of the NG9-1-1 system is an open architecture and relies on many different technical standards to support its requirements. To date, many communications, networking, and telephony standards that will affect NG9-1-1 are still in development or need to be selected and accepted before the NG9-1-1 system can be implemented. Until NG9-1-1 related baseline standards are developed, selected, fully vetted, and ultimately adopted, uncertainty among 9-1-1 decision makers and service and equipment providers may hinder the migration to NG9-1-1.

NG9-1-1 System Access and Security Controls. The NG9-1-1 system, like other mission-critical systems, will continually encounter attempts at illegal access, including concerted malicious attacks (e.g., denial of service, virus or worm transmission, etc.). Therefore, to mitigate security risks and control access to the NG9-1-1 system, security controls and certification and authentication mechanisms need to be developed to identify and determine the access methods, rules, and controls by which users and systems are allowed to access the NG9-1-1 network in a prescribed and standardized manner.

Location Determination and Acquisition for NG9-1-1. A 9-1-1 caller's location information is needed to route emergency calls to the most appropriate

PSAP. Subsequently, the call taker must identify the location of the emergency to ensure that responder agencies are dispatched to the correct location. In the NG9-1-1 environment, the increased number of devices that can call 9-1-1 and associated services, infrastructure, and access providers that provide the means will make it challenging to determine how to acquire the location information and who is responsible for providing that information.

NG9-1-1 Call Routing and Prioritization Protocol and Business Rules Development. In the NG9-1-1 environment, call routing and prioritization will be challenging because the open architecture of the NG9-1-1 system will enable routing of 9-1-1 calls, based on multiple factors, throughout the Nation and internationally instead of being limited to a specific jurisdiction. It is expected that NG9-1-1 calls will continue to be routed based, to a degree, on location; however, it has not been determined what other factors should be used to route and prioritize NG9-1-1 calls.

NG9-1-1 Key Architecture Considerations. The supporting technology infrastructure for NG9-1-1 must meet a variety of needs of the global public safety communications professionals. Systems must be scalable, both vertically and horizontally, as well as extensible, permitting future technologies to be woven into the system without requiring wholesale replacement. Ensuring interoperability across disparate systems and eliminating single points of failure to ensure reliability are of utmost concern.

What Has Been Done?

Recently, some progress has been made to partially resolve elements of the NG9-1-1 standards and technology issues:

SDO Standards Efforts. Several SDOs, including Internet Engineering Task Force (IETF), NENA, and APCO, have been actively working to develop NG9-1-1 related standards. For example, NENA's i3 Technical Requirements Document identifies standards and requirements for various components of the NG9-1-1 System Architecture. Also, IETF Emergency Context Resolution with Internet Technologies (ECRIT) has developed the Location to Service Translation Protocol (LoST) to serve as the protocol to route NG9-1-1 calls to the most appropriate PSAP. In addition, the NG9-1-1 relies on an IP-based infrastructure that allows the transmission of voice, video, images, and data using different protocols. For example, voice calls will use the Session Initiation Protocol (SIP) which is the focus of one of the IETF's working groups. The OASIS Emergency Management Technical Committee currently promotes data interoperability standards and several organizations are collaborating on a suite of specifications called the Emergency Data Exchange Language (EDXL). EDXL is an Extensible Markup Language (XML)-based model intended to create an integrated framework for a wide range of data exchange standards to support emergency operations.[23]

NG9-1-1 Initiative Architecture Analysis Report. As part of the USDOT NG9-1-1 Initiative, an analysis of the notional NG9-1-1 architecture was performed. This report[24] outlines a high-level analysis and identification of the technological deficiencies and gaps that must be addressed in order to achieve a nationwide solution. Although not a blueprint for implementing a production system, the analysis provides the framework for how a system could be constructed and the components and subsystems that must be considered.

Federal Legislative Mandate for IP-Enabled Emergency Network and Location Information and Delivery. The U.S. Senate has passed "S.428 *IP-Enabled Voice Communications and Public Safety Act of 2007,[25]"* legislation that makes clear that it supports the migration to a national IP-enabled emergency network capable of receiving and responding to emergency communications and improving information sharing among emergency responders. Furthermore, it states that the FCC shall coordinate with public safety organizations, industry participants, and others to develop standards and best practices, and promote consistency for location information.

What Could Be Done to Address NG9-1-1 Standards and Technology?

Additional options are available throughout the 9-1-1 community to continue to address potential standards and technology shortfalls that could affect nationwide migration to NG9-1-1. These options include potential strategies or paths that would—

Complete and accept NG9-1-1 open standards and understand future technology trends to encourage system interoperability and emergency data sharing

Establish system access and security controls to protect and manage access to the NG9-1-1 system of systems

Determine the responsible entity and mechanisms for location acquisition and determination

Determine routing and prioritization protocols and business rules.

Complete and Accept NG9-1-1 Open Standards and Understand Future Technology Trends to Encourage System Interoperability and Emergency Datasharing

The NG9-1-1 system relies on open, non-proprietary standards to develop an architectural framework that promotes system interoperability and emergency data sharing. Without the development and acceptance of NG9-1-1 related standards by the 9-1-1 community and service providers, the ability to interconnect systems and achieve the NG9-1-1 vision to provide ubiquitous, interconnected NG9-1-1 services across the Nation may not be realized. The Federal Government and 9-1-1 community should participate in the identification and development of critical and necessary NG9-1-1 related standards to ensure the standards meet the needs of the 9-1-1 community. Furthermore, the Federal Government and SDOs should coordinate efforts to establish a certification program to ensure that equipment and services provided for NG9-1-1 meet the open standards defined by the industry-accepted architectural model. Finally, the Federal Government could identify a standard coordinating entity and apply funding to support dedicated attention to the development of standards and technologies considered to be essential to NG9-1-1 and coordinate emergency data standard coordination and harmonization. Table 9 outlines some of the options federal and state governments, and SDOs can take to develop and promote acceptance of NG9-1-1 open standards.

Table 9. Options to Develop and Promote Acceptance of NG9-1-1 Open Standards

• Responsible Party	• Options
• Federal Government 	• Coordinate with 9-1-1 stakeholders (e.g., SDOs, private and public stakeholders) to identify all standards work and technology development currently underway within relevant communications.
	• Coordinate with SDOs to conduct a gap analysis of existing standards development work against the industry-accepted architectural model and identify standards that will need to be developed and/or are already developed standards that are inconsistent with the model.
	• Identify and encourage appropriate SDOs and industry groups to address standards work that must be completed or modified to ensure a comprehensive standards and technology environment for NG9-1-1.

	• Identify and leadership/coordinating entity and apply funding to support dedicated attention to the development of standards and technologies considered essential to NG9-1-1 and coordinate emergency data standard coordination and harmonization.
	• Encourage states to enact laws or regulations that mandate the use of open standards for NG9-1-1 systems.
	• Assist in the development of a certification program to ensure that NG9-1-1 services and equipment meet the open standards defined by the industry-accepted architectural model.
• State Government 	• Require 9-1-1 Authorities to procure equipment that has been certified to meet the open standards requirements
• SDOs 	• Work with the 9-1-1 community to identify critical standards, future technologies, and all relevant data generators and user communities based on an industry-accepted architectural model and detailed requirements.
	• Conduct a gap analysis of existing standards development work against the industry-accepted architectural model and identify standards that will need to be developed and/or already developed standards that are inconsistent with the model.
	• Develop a certification program to encourage and verify that NG9-1-1 services and equipment meet the open standards defined by the industry-accepted architectural model.
• Responsible Party	• Options
•	• Strengthen rules and procedures to ensure that NG9-1-1 essential standards and technology development occur in an open, fair, and competitively neutral environment (recognizing the nature of technology convergence, and competitive interests involved).
• Service and Equipment Providers 	• Participate in the definition and development of open standards for the NG9-1-1 System and equipment.
	• Develop NG9-1-1 equipment and update telecommunications services to meet the open standards identified by the 9-1-1 community and federal government to promote interoperability.
	• Participate in certification of equipment to ensure the equipment meets the open standard requirements set forth by the 9-1-1 community and federal government.

Table 10. Options to Establish Security Controls and Access Rights

• Responsible Party	• Options
• Federal Government 	• Consider initiating establishment of new security regulations or modification of existing security regulations to promote consistency among states.
	• Identify and modify, as necessary, federal legislative or regulatory provisions that may constrain the transition to the NG9-1-1 environment, thereby enabling unbundled service offerings.
	• Adopt a national-level certification and authentication process determined by the SDOs to ensure service providers and 9-1-1 Authorities meet the security and system access requirements.
• Responsible Party	• Options
	• Identify and leverage industry and defense and homeland security solutions, standards, and best practices to ensure a consistent and appropriate level of security throughout the interconnected environment of NG9-1-1.
	• Promote the development of identity and data rights access and management to control access to the NG9-1-1 system and data.
• State Utility Commissions 	• Coordinate and modify existing state regulations to allow service providers beyond the common carriers to access the NG9-1-1 system.
	• Identify and modify, as necessary, state contractual provisions that may constrain the transition to the NG9-1-1 environment.
	• Assist SDOs in developing a certification and authentication process to ensure service providers and 9-1-1 Authorities meet security and system access requirements.
• SDOs 	• Develop a certification and authentication process to encourage service providers and 9-1-1 Authorities meet the security and system access requirements.
	• Identify and leverage defense and homeland security solutions, standards, and best practices to establish security and system access standards and requirements for NG9-1-1.
	• Define identity and data rights access and management standards that can ensure only specific entities or individuals may access the NG9-1-1 system and data.

Establish System Access and Security Controls to Protect and Manage Access to the NG9-1-1 System of Systems

The security of and authorized access to the NG9-1-1 system is critical to ensuring that the NG9-1-1 system of systems is secure from security breaches and illegal users to prevent disruption of the delivery of a 9-1-1 call and public safety response to emergencies. The Federal Government, state utility commissions, and SDOs will play a major role in ensuring that 9-1-1 Authorities and service providers have standards for meeting the security controls and system access requirements. Therefore, the Federal Government should identify and leverage defense and homeland security solutions, standards, and best practices to determine an appropriate level of security for the NG9-1-1 system. In addition, the Federal Government should work with the state utility commissions to modify existing regulations that limit access to the 9-1-1 system and determine a certification and authentication process to ensure service providers and 9-1-1 Authorities meet security and system access requirements. Table 10 outlines some of the options Federal Government, state utility commissions, and SDOs can take to establish and promote security controls and system access rights.

Determine Responsible Entity and Mechanisms for Location Determination and Acquisition

As new devices become available that are capable of calling 9-1-1, challenges will arise to determine and acquire the location of the caller and device, especially if the device is transportable. Furthermore, as new types of service providers (e.g., infrastructure, access) enter the market, it will be a challenge to identify the entity that would be responsible for providing the location information. Similar to efforts requiring cellular telephone and VoIP service providers to obtain and provide location for 9-1-1 calls from cellular and VoIP telephones, the Federal Government should consider identifying and mandating specific service providers to collect accurate location information. In addition to collecting location information, the information should be standardized to promote the ability to easily share data among PSAPs and emergency response entities. Table 11 outlines some of the options Federal Government and SDOs can take to determine the responsible entity and mechanisms for location determination and acquisition.

Table 11. Options to Determine Responsible Entity and Mechanisms for Location Determination and Acquisition

• Responsible Party	• Options
• Federal Government 	• Develop a national mandate that requires service providers (access and transport network and devices) to meet specific standards or requirements for obtaining and providing accurate location information for all call types that access NG9-1-1.
	• Determine the responsibility for location validation and management of national and/or regional databases necessary for NG9-1-1.
	• Examine issue of who within the service provider community will be responsible for providing accurate location information for NG9-1-1 calls.

• State Government 	• Determine the responsibility for location validation and management of national, regional, and/or local databases necessary for NG9-1-1.
	• Determine who within the service provider community will be responsible for providing accurate location information for NG9-1-1 calls within state..
• SDOs 	• Develop data standardization for the delivery of location information whether it is civic or geospatial.
• Service and Equipment Providers 	• Develop methods or modify existing methods to obtain the necessary location information for all call types that access NG9-1-1 and provide as mandated by the federal government.

Determine Routing and Prioritization Protocols and Business Rules

Accurately routing 9-1-1 calls to the appropriate PSAP will be more complex in the NG9-1-1 environment because of the potential amount of data that can be used to route and prioritize NG9-1-1 calls. To ensure that the protocols and business rules developed for routing meet the needs of the 9-1-1 community, SDOs should collaborate with the 9-1-1 community and emergency response entities to identify what information is important and how the call should be routed. Table 12 outlines the options for developing routing and prioritization protocols and business rules.

Table 12. Options to Determine Routing and Prioritization Protocols and Business Rules

• Responsible Party	• Options
• SDOs 	• Identify from the user community what types of data are needed and standardize the types of data, decisions, and/or policies that should be used to route and prioritize NG9-1-1 calls.
	• Develop nationally defined protocols and business rules and options (either decision and/or policy based).

Does It Matter?

It is imperative that the necessary NG9-1-1 related standards and technology are determined and available for the 9-1-1 Authorities and PSAPs to begin transitioning to an open, non-proprietary NG9-1-1 system. Without the critical standards and technologies in place, service and equipment providers may develop new, non-proprietary solutions that will not benefit the global community. This can affect the ability of PSAPs and emergency response entities to effectively share information and be interoperable. Furthermore, without the critical processes and protocols (e.g., certification and authentication, routing business rules), the benefits of the NG9-1-1 system, including routing based on criteria beyond location and allowing service providers beyond common carriers to connect to the 9-1-1 system, may not be realized.

For Additional Information

As the NG9-1-1 system begins to become a reality, additional information regarding standards and technology can be found at:

USDOT Research and Innovative Technology Administration (RITA) ITS Next Generation 9-1-1 Initiative

USDOT NHTSA / DOC NTIA National 9-1-1 ICO

NENA TDC and Technical Committee

Network Reliability and Interoperability Council (NRIC) VII Focus Group 1: Enhanced 911

Internet Engineering Task Force (IETF)

o Emergency Context Resolution with Internet Technologies (ECRIT)

Geographic Location/Privacy (GEOPRIV)

Oasis – Emergency Management

STRATEGIC OPTIONS FOR NG9-1-1 GOVERNANCE AND POLICY

Background

The current 9-1-1 environment is a structured process with relatively clear delineation regarding roles and responsibilities among the parties as defined by common practices and statutes. As NG9-1-1 deployment begins, these relationships will become less clear, and the existing legal and regulatory environment will not accommodate new technologies and arrangements. The deployment of NG9-1-1 will most likely increase coordination and partnerships among government stakeholders, 9-1-1 Authorities, and PSAP Administrators in planning and implementing NG9-1-1. Coordination with industry and the general public will also be important to address stakeholder concerns and to manage expectations. As a result, legislative and regulatory arrangements and demarcation points at every level of government will need to be reexamined and likely modified to effectively deploy NG9-1-1. The corresponding NG9-1-1 issues that should be addressed include the following:

NG9-1-1 Roles and Responsibilities. In an NG9-1-1 environment, the roles of the PSAPs, responders, and related entities are expected to expand beyond traditional 9-1-1 services with higher levels of interaction, managed situational intelligence, enhanced capabilities, and more comprehensive communication and coordinated response. Similarly, the relative roles of local, state, and federal government in guiding the evolution and migration of the current 9-1-1 system to NG9-1-1 are unclear. Lack of coordination and clarity regarding these roles will delay the decision-making needed to continue progress toward the national implementation of NG9-1-1 and may delay the process or potentially impair its effectiveness.

NG9-1-1 Legislative and Regulatory Barriers. While 9-1-1 requirements are established at the federal level for wireless and VoIP services, there is an important state and local role in determining service arrangements, costs, and funding mechanisms. The appropriate level of government oversight and regulation for IP-based communications for 9-1-1 services requires ongoing policy-making processes. Making changes in the service delivery environment may require commensurate changes in legislative and regulatory policy.

NG9-1-1 Privacy Concerns. The NG9-1-1 system allows the transmission and storage of a variety of personal information, such as medical and health data, photographs, and other visual images. The range of personal and identifiable information that can be collected and transmitted by NG9-1-1 will likely raise concerns from the public and interest groups tracking privacy issues.

Liability Protection for Future Services. In the current system, 9-1-1 wireless and wireline service providers are afforded statutory protection from liability. NG9-1-1 will foster new types of services, along with their providers. Their standing with respect to liability has not been defined by law, or is unclear. This may discourage the entry of new service providers and innovative data that could result in more effective emergency response services.

NG9-1-1 Awareness and Accessibility. As NG9-1-1 systems are deployed, the general public will want to know what, how, and when next generation services will be available in their area. The public will likely have questions regarding NG9-1-1 system capabilities and limitations. The government may need to examine the impact NG9-1-1 deployment has on the elderly, deaf and hard of hearing, disabled, and non-English speaking populations. Consequently, keeping the public informed and involved throughout the planning and deployment of NG9-1-1 will be important to its ultimate success.

What Has Been Done?

Recent progress has been made on addressing several NG9-1-1 governance and policy issues:

Defining Responsibilities and Accountability in NG9-1-1. Although federal, state, and local government roles and responsibilities still need to be defined, organizations like the 9-1-1 ICO and DHS have begun to consider 9-1-1 and NG9-1-1 within the context of homeland security policy and nationwide emergency communication planning.

Federal Legislation. The Congress is considering legislation—the IP-Enabled Voice Communications and Public Safety Act of 2007—that provides liability parity for PSAPs and for VoIP providers and their third-party vendors equivalent to what is provided for wireline and wireless service providers. This legislation also confirms state authority to impose and collect 9-1-1 fees on IP-enabled services.

Education and Awareness. A resolution has been introduced in Congress that would support the designation of a month each year as "National 9-1-1 Education Month." National 9-1-1 Education Month would promote 9-1-1 education—which can include awareness of NG9-1-1— and be a tool to help support the efforts of those currently involved in 9-1-1 education.

What Could Be Done to Address NG9-1-1 Governance and PolicyIssues?

Additional options are available at all levels of government to address governance and policy that could affect nationwide migration to NG9-1-1. These options include potential strategies or paths that would—

Clarify jurisdictional frameworks and responsibilities and identify the coordination required at each level of government to enable NG9-1-1

Update regulations, legislation, and other policies to—

- o Reflect communications capabilities in an NG9-1-1 system
- o Protect the confidentiality of new types of personally identifiable information (PII)
- o Extend liability protection and parity for future services connecting to 9-1-1

Ensure continued access to 9-1-1 system on current and future devices and services in which users reasonably expects to access to 9-1-1.

Clarify Jurisdictional Frameworks, Responsibilities, and Coordination Required at All Levels of Government to Enable NG9-1-1

The Nation's experience with the rollout of wireless E9-1-1 services has demonstrated that lack of coordination among the many public and private entities engaged in implementation delays deployment. When compared with wireless E9-1-1 implementation, the rollout of NG9-1-1 will be more complex because of new types of services and service providers and changing roles of existing 9-1-1 system stakeholders. Therefore, federal, state, and local government leadership roles will need to be further defined to facilitate the deployment and operation of the NG9-1-1 system.

Coordinated approaches and partnerships among 9-1-1 Authorities, PSAPs, emergency responders, and government officials (elected or appointed) will be essential to take full advantage of the next generation opportunities and capabilities. At the federal level, policymakers could further empower entities like the 9-1-1 ICO with support from other federal entities such as DHS to coordinate and facilitate NG9-1-1 planning nationwide. Defining roles and responsibilities for regional and statewide coordination will also be important to provide accountability and guidance. Table 13 outlines the options for clarifying jurisdictional frameworks and responsibilities and identifying the coordination required at each level of government to enable NG9-1-1.

Table 13. Options to Clarify Responsibilities and Identify Required Coordination for NG9-1-1

• Responsible Party	• Options
• Federal Government 	• Identify a lead agency or central sponsor for guidance and coordination of NG9-1-1 at the national level.
	• Strongly encourage states to convene their state 9-1-1 Administrator, state homeland security/emergency management director, and state chief information officer (CIO) to plan for establishment of NG9-1-1 and its underlying emergency services inter-networks
	• Consider modifying current legislation to assign state/regional 9-1-1 Authorities the responsibility and authority to transition to and manage the NG9-1-1 system.
	• Facilitate appropriate regional or national working groups and forums at all levels to coordinate business rules and data rights management with emergency systems at the state and local levels.
• State Government 	• Convene appropriate state level officials (e.g., state 9-1-1 Administrator, state homeland security/emergency management director, and state CIO) to plan for establishment of NG9-1-1 and its underlying emergency services internetwork.
	• Designate a central coordinating body and/or mechanism for NG9-1-1 implementation for each state (or region) where one does not already exist.
	• Develop a statewide NG9-1-1 implementation plan based on effective stakeholder involvement, including identifying responsible parties, goals, and milestones.
	• Consider modifying current legislation to assign state 9-1-1 Authorities with the responsibility and authority to transition to and manage the NG9-1-1 system.
	• Consider modifying current legislation to assign public utility commissions the authority and responsibility for determining certification requirements for telecommunications services and service providers that may access NG9-1-1 as appropriate at the state level.
	• Coordinate business rules and data rights management at all levels of stakeholders.

• 9-1-1Authorities 	• Participate in the development of a statewide NG9-1-1 implementation plan, including identifying responsible parties, goals, and milestones.
	• Work with federal, state, and local level entities to coordinate business rules and data rights management within the appropriate jurisdictional boundaries.

Update regulations, legislation, and other policies to—

Reflect communications capabilities in an NG9-1-1 system
Protect the confidentiality of new types of PII
Extend liability protection and parity for PSAPs and new services

Typically, 9-1-1 telecommunications service providers and the services they offer are regulated through their state public utility commission and the FCC. Because current 9-1-1 services are based on regulated telecommunications companies, new non-Local Exchange Carrier (LEC) service providers with new technologies may be unable to participate in 9-1-1 service delivery under current rules. Although the FCC has mandated that the wireless carriers and VoIP service providers have access to the 9-1-1 network, the current 9-1-1 governance environment does not always allow access for new, advanced communications technologies. Furthermore, in the current regulatory environment, 9-1-1 service providers have insufficient incentive to fund and deploy advanced system architecture.[26] There are a number of legislative options that policymakers could use to address regulatory shortcomings.

Similarly, there are currently inconsistent policies concerning the disclosure of customer-specific information by telecommunications providers to government agencies for the delivery of emergency services. The different legal policies have resulted in delays in providing emergency response during real emergencies and can adversely affect effective 9-1-1 database provisioning. At the federal and state level, legislators should examine these issues and make consistent the relevant statutory provisions on how to handle and treat new and expanded personal information available with NG9-1-1.

Finally, liability protection will be needed for communications services that seek to connect to NG9-1-1 but are not yet required by the FCC to provide 9-1-1/NG9-1-1. Based on pending legislation, this could occur with approval from the appropriate state or local 9-1-1 Authority. To ensure that future service providers are treated equally, the Federal Government should explore what criteria authorities should use in granting approval. Table 14 outlines the options for reducing potential regulatory and legislative barriers for NG9-1-1.

Table 14. Options to Reduce Potential Regulatory and Legislative Barriers for NG9-1-1

• Responsible Party	• Options
• Federal Government 	• Update federal laws and encourage the modification of state laws that may prohibit the use of location information for call routing beyond those covered under traditional 9-1-1 (e.g., include appropriate N-1-1 entities and 800-type emergency provisioning, such as poison control centers and suicide hotlines)
	• Develop new or modify existing federal laws to address confidentiality of new data sources and types
	• Determine whether existing liability protection for PSAPs and service providers will continue to apply in the NG9-1-1 environment
	• Approve and extend, if necessary, liability protection and parity for PSAPs and new service providers
	• Review existing laws and determine whether the existing law addresses the confidentiality of new data sources and types (e.g., medical) that could be considered PII
	• Develop new or modify existing state laws to address confidentiality of new data sources and types
	• Consider updating laws that prohibit the transmission of non-human initiated calls to a PSAP absent human intervention to accommodate sensor alerts from verified sources
	• Update state laws that prohibit 9-1-1 service from being provided by entities other than tariffed LECs/Competitive Local Exchange Carrier (CLEC)
	• Review and update laws to ensure existing statutes do not prohibit use of "virtual PSAPs."
	• Update state laws that may prohibit the use of customer/device location information for routing of other calls beyond those covered under traditional 9-1-1 (e.g., include appropriate N-1-1 entities and 800-type emergency provisioning, such as poison control centers and suicide hotlines).
	• Determine whether existing liability protection for PSAPs and service providers will continue to apply in the NG9-1-1 environment.
	• Approve and extend, if necessary, liability protection and parity for PSAPs and new service providers.

Ensure Continued Access to 9-1-1 System on Current and Future Devices and Services in which Users Reasonably Expect to Access to 9-1-1

Public awareness has always been important to 9-1-1, and most 9-1-1 programs include a public education component. As text messaging and the use of camera telephones increases, expectations and misperceptions of what can and cannot be delivered to a 9-1-1 call taker increases. As these services continue to become more mainstream, more and more consumers will be frustrated and confused about why the capabilities they use daily do not work when they need to call 9-1-1. The public expects 9-1-1 centers to keep up with technological change and that the NG9-1-1 System will deliver the same level of service to which people have become accustomed, if not a higher level of service. Therefore, the question arises: Should public expectations regarding the accuracy and speed of call delivery and location determination be lowered? As NG9-1-1 services are rolled out and implemented, a key issue will be to ensure that the public understands how, where, and when next generation services are available.

In addition, NG9-1-1 offers new communications options for the elderly, deaf and hard of hearing, disabled, and non-English speaking populations that may need to be addressed by policymakers. Many deaf and hard of hearing persons are replacing their legacy communications devices (e.g., TTYs) and using mainstream text communications devices, IP-Relay Services and Video Remote Interpreting (VRI) Services. NG9-1-1 is expected to eventually allow direct connection between these services and 9-1-1 and reduce delays in reaching call-takers. As NG9-1-1 deployments allow this functionality, regulations ensuring that deaf and hard of hearing individuals can be efficiently connected to 9-1-1 must be immediately implemented. Table 15 outlines the options for improving awareness and accessibility of NG9-1-1.

Table 15. Options to Improve Awareness and Accessibility of NG9-1-1

• Responsible Party	• Options
• Federal Government 	• Consider regulations ensuring that deaf and hard of hearing individuals can be efficiently connected to 9-1-1 when such NG9-1-1 capability is available.
	• Provide clarity and update on federal regulatory boundaries and authority.
	• Educate the public and policymakers on the availability of NG9-1-1 capabilities.
	• Consider additional legislative, regulatory, of financial incentive to ensure that states and authorities provide individuals with disabilities access to the 9-1-1 system using the current and future technology and

	devices that are used for daily communications.
State Government 	• Educate the public and policymakers on the availability of NG9-1-1 capabilities.
9-1-1 Authorities 	• Educate the public and policymakers on the availability of NG9-1-1 capabilities.
	• Educate the public and policymakers on the availability of NG9-1-1 capabilities on telecommunications devices, as well as services used (e.g., VoIP). Similar to VoIP providers informing customers of the limitations of 9-1-1 capabilities with VoIP service.
General Public 	• Understand and be aware of available NG9-1-1 capabilities and coverage.
	• Provide recommendations and input into the NG9-1-1 development and deployment process to account for community needs.

Why Does It Matter?

Call takers and emergency responders will need to have the appropriate governance, policies, and mechanisms to provide a coordinated and appropriate response in an NG9-1-1 environment. In many situations, the lack of coordination and necessary policies could impede the ability of PSAPs and call takers to respond quickly and effectively, or to their greatest capability. Elected and appointed officials, senior government executives, and communications managers must foster and support effective NG9-1-1 partnerships and the appropriate statutory and regulatory policies. They must also ensure that the general public is part of the NG9-1-1 deployment process. The NG9-1-1 environment allows for and promotes a higher level of coordination among federal, state, and local government and the first responder community, which is limited with today's 9-1-1 system.

For Additional Information

Additional information regarding governance and policy can be found at:

NENA: 2008 9-1-1 Goes to Washington
NENA: Next Generation Partner Program Policy Issues Update
NENA: Government Affairs Update
E911 Institute

CONCLUSION

The completion of the USDOT NG9-1-1 Initiative marks a critical transition point for the public, emergency communications, and public safety communities. Stakeholders are increasingly concerned about the evolution and use of new technologies by the general public and emergency responders. The NG9-1-1 Initiative has established a substantial body of knowledge on the vital transition issues and associated strategic options that the Nation can implement to address NG9-1-1 transition issues. Through an extensive series of multi-stakeholder workshops, the Initiative identified a broad range of perspectives on the role of the emergency response community, industry, decision-makers, and the general public, as well as some important constraints on the NG9-1-1 transition.

However, limited time and resources mean that not every stakeholder community could engage as intensely in some stages of this process. Furthermore, the restrictions of the project mean that USDOT will not be able to address a wider range of the general public's perceptions and concerns about new technologies and NG9-1-1. Consequently, it should be noted that there are undoubtedly shortcomings within this plan that will be addressed based on stakeholder feedback and input in the final *NG9-1-1 Transition Plan*. It is important to recognize that this *NG9-1-1 Preliminary Transition Plan* is the beginning of a process—not the end.

The commitment demonstrated by diverse stakeholders over the past 2 years and the energy generated on behalf of the NG9-1-1 Initiative needs to be channeled toward lasting, tangible progress. There is an expectation that the period following the NG9-1-1 Initiative will be marked by action rather than words. Therefore, over the coming months, the NG9-1-1 Initiative will use the *NG9-1-1 Preliminary Transition Plan* to ensure the dialogue and cooperation established by the initiative is continued and extended in the future. All groups and individuals with a stake in NG9-1-1—industry, the general public, governments, communities, and nongovernmental organizations—have a right and a responsibility to have their voices heard in dialogue about the future of 9-1-1. The NG9-1-1 initiative will build on and expand its *Preliminary Transition Plan* throughout 2008, based on the feedback it receives after this document is published and released to the public. At the end of 2008, the NG9-1-1 Initiative will publish a final *NG9-1-1 Transition Plan* that serves as an agenda for action and a foundation for the 9-1-1 community in planning and deploying NG9-1-1.

APPENDIX A: NG9-1-1 TRANSITION ISSUES REPORT

Introduction

A key outcome of the U.S. Department of Transportation (USDOT) Next Generation 9-1-1 (NG9-1-1) Initiative is the identification of well-supported options and crucial transition issues for migrating from today's 9-1-1 system to a NG9-1-1 system across the country. The following analysis documents and summarizes many of the issues facing government, industry, and the general public as the Nation plans for future NG9-1-1 implementation. The scope and breadth of the transition issues characterized here are intentionally broad and includes issues that need to be addressed at the national, state, regional, and local levels to implement NG9-1-1. Based on stakeholder input and existing research, this analysis explores several of the key issues that need to be addressed for successful transition to the NG9-1-1 environment by posing the following questions:

What are the key questions that need to be answered and what needs to be decided?
What is the impact of each issue on NG9-1-1 transition activities?
Who is affected by the issues and how?
How will varying deployment scenarios change the nature of an issue?
What are the assumptions or constraints for each issue?

This analysis recognizes that each issue may affect transition to NG9-1-1 and its stakeholders in different ways. Similarly, the priority of the key issues is also expected to vary over time and by location. Therefore, this analysis presents a list of transition issues that is ordered for ease of reference, not in order of priority. Lastly, as highlighted in this analysis, the answers to these critical questions acknowledge that the migration path chosen by a state, regional, or local 9-1-1 Authority[27] may lessen or exacerbate the importance or impact of these transition issues.

Deployment Framework

As defined by the NG9-1-1 Initiative, the NG9-1-1 System is expected to be an interconnected system of local and regional emergency services networks ("system of systems").[28] However, the boundaries of emergency service networks may vary, depending on local requirements and organizational frameworks. Consequently, although the architecture and underlying requirements of NG9-1-1 will be consistent, it is expected that there will be several paths to implementation of NG9-1-1—all with merit. In some cases, the path to NG9-1-1 implementation will depend on the underlying infrastructure and state of the Public Service Answering Points (PSAP) and 9-1-1 Authorities. Regardless of where PSAPs and 9-1-1 Authorities begin, the majority of NG9-1-1 implementation paths will likely be characterized by some variation of the two deployment frameworks discussed below and in table 1:

Independent Deployment: Building out NG9-1-1 in a piecemeal or "stand-alone" approach that local 9-1-1 Authorities (e.g., townships, counties, parishes, cities) could pursue by deploying NG9-1-1 capabilities without wider coordination or interconnecting guidance and planning. These local 9-1-1 Authorities would have the flexibility to determine how and when to implement the NG9-1-1 requirements and standards. Potentially, different PSAPs would continue to have inconsistent or incomplete 9-1-1 capabilities and functionality, and the opportunity for sharing costs, infrastructure and services with other 9-1-1 Authorities on a larger scale would be limited.

Coordinated Deployment: Building out NG9-1-1 in a uniformed, guided approach by regional or state 9-1-1 Authorities to upgrade identified PSAPs to meet a consistent set of nationally accepted NG9-1-1 requirements and standards. PSAPs would have coordinated functions and interconnection. To reach large-scale deployment, 9-1-1 Authorities could employ several different transition paths, including upgrading individual PSAPs, upgrading PSAPs within a region or state, providing functionally specialized PSAPs, consolidating PSAPs, and/or phased approaches based on size or resource availability—but all would involve guidance and coordination through appropriate regional, state, and national mechanisms.

Table 1. Summary of Deployment Frameworks

•	• Independent	• Coordinated
• Summary	• Customized set of capabilities reflecting conditions, needs, and priorities at the local level	• Consistent, guided standardized capabilities and technology platforms at the regional or state level
• How	• Independent upgrade of individual PSAPs or some PSAPs within a local jurisdiction (e.g., township, county, parish, city) • Inconsistent capabilities	• Centrally planned and coordinated upgrade of individual PSAPs, PSAPs within a region, or PSAPs within a state • Coordinated phased approach to upgrade the PSAPs (e.g., based on size of population served and/or availability of resources) • Consistent capabilities
• Who Defines	• Individual 9-1-1 Authorities	• Coordinated 9-1-1 Authorities
• Advantages	• Enables migration to NG9-1-1 based upon local conditions, needs, and priorities	• Promotes consistent and coordinated adoption of NG9-1-1 capabilities and functions • Potentially enables ubiquitous NG9-1-1 capability throughout the regions and states involved
• Disadvantages	• Limits the opportunity for cost and resource sharing, as well as for consistent large-scale service delivery • Promotes a fractured community further dividing "haves" and "have nots" • Creates limits in realizing full benefit of NG9-1-1 capabilities	• Requires a high level of coordination, cooperation, and planning among the state, regional, and local 9-1-1 Authorities involved • May encounter limited support from state or regional entities in some instances

NG9-1-1 Transition Issues

Issue: Current Funding, Budgeting, and Cost Recovery Policies May Not Be Able to Support the Implementation and Sustainment of NG9-1-1

Impact on NG9-1-1: This may create an inconsistent nationwide footprint of NG9-1-1 by furthering the disparity among PSAPs (e.g., between the "haves" and "have nots"), as well as delay the implementation of NG9-1-1 throughout the Nation. Inconsistent application of NG9-1-1 features will allow persistent misconception by end users who currently expect that the same level and quality of service is available throughout the Nation.

Stakeholder(s) Affected: Funding issues will affect a variety of stakeholders, including 9-1-1 Authorities, state and local governments. 9-1-1 Authorities require sufficient financial resources to upgrade to and maintain the NG9-1-1 system. State and local government agencies may need to determine how to adapt the 9-1-1 surcharge model to accommodate how 9-1-1 will work in tomorrow's environment to ensure that sufficient funds are collected to maintain the 9-1-1 services. In addition, states and local entities that are expecting to use funds designated for 9-1-1, may continue to face funding shortfalls if such funds are redirected to other programs.

Background and Description: The current and prevailing 9-1-1 funding model is to assess subscriber fees on wireline, wireless, and some Voice over Internet Protocol (VoIP) telephone services, which are collected by service providers and then transferred to government agencies to support 9-1-1 services. There is little consistency among and within states on 9-1-1 surcharge rates, which often differ based on the service type (e.g., wireless, wireline, or VoIP) or the location in which the fee is being collected. 9-1-1 Authorities at every level are already experiencing a decline in wireline surcharge revenues as consumers abandon their wireline services and move to using wireless and IP-based technologies (e.g., VoIP) for their primary voice communications. Furthermore, many states and local governments are continuing to divert 9-1-1 funds for non-9-1-1 uses such as balancing state budgets or funding other initiatives. The current funding allocation and cost recovery mechanisms may be insufficient to ensure sustainable funding for future 9-1-1 services, and raises issues of fairness and equitability. For example, there may be new types of equipment that may generate additional costs, such as devices for IP routing and automatic call distribution (ACD) that could be shared among State and local jurisdictions and are not eligible for payment by the funding allocation and cost recovery mechanisms. In addition, sharing the NG9-1-1 backbone network among many state and local jurisdictions will make it difficult to determine how much each jurisdiction must contribute to total network costs.

Assumptions/Considerations: The analysis of this issue assumes funding shortfalls are generally contingent on the continued trend and transition away from traditional wireline service and toward wireless and VoIP services. In addition, it is assumed that there is a failure of state legislatures to take remedial measures to update state laws.

Issue: Security Controls for the NG9-1-1 System and Emergency Data Need to Be Defined and Properly Managed and Maintained

Impact on NG9-1-1: The NG9-1-1 network, like other networks, will continually encounter attempts at illegal access including concerted attacks (e.g., denial of service). If the NG9-1-1 network is compromised by security breaches, it is possible that there will be a disruption in delivering a 9-1-1 call from the public to a PSAP, resulting in delayed or complete interruption of public safety response to emergencies.

Stakeholder(s) Affected: Security is an overarching concern for the entire NG9-1-1 system and will affect 9-1-1 stakeholders, including service providers, equipment providers, 9-1-1 Authorities, and PSAPs. For example, service providers may be responsible for ensuring that the location databases (e.g., Location Information Server [LIS], Location-to-Service Translation [LoST]) and access networks they maintain are secure against threats. 9-1-1 Authorities, on the other hand, may be responsible for ensuring that all connection points to the PSAPs from the NG9-1-1 network meet or exceed the security controls needed to protect the system from hackers. PSAPs are increasingly putting their public safety systems on a shared city or county network, further exposing their systems to attacks from outside sources.

Background and Description: Security control issues for the NG9-1-1 network can affect several different areas of the network, including connection or gateway points to the PSAPs, to the IP access network, to the Internet, and to other public networks via the Emergency Services Network (ESNet), and to external databases potentially managed by other entities. The NG9-1-1 network architecture is designed to interface with various legacy and next generation access and emergency service networks to provide a seamless, interconnected delivery and response environment. The network connection or gateway points between the NG9-1-1 network and other networks or external databases may be exposed to security vulnerabilities, including denial of service attacks, spoofing, and malformed messages that can disrupt 9-1-1 services (e.g., flooding the NG9-1-1 system with fake information and preventing real 9-1-1 calls from being completed). In addition, interfaces may be created with other networks, including public safety entities, third party service providers, and public Internet, which may expose the NG9-1-1 Network to viruses or hacker attacks. Likewise, databases used by the NG9-1-1 network and managed by other entities need to be secured to prevent hackers from potentially manipulating the information in the database. To ensure the NG9-1-1 system of systems is secure, it is crucial that security controls are established to mitigate security risks throughout the NG9-1-1 network.

Security controls throughout the NG9-1-1 network may need to be tailored for each network connection because security needs may vary.

Assumptions/Considerations: The analysis of this issue assumes that NG9-1-1 implementations will incorporate industry-leading best practices for security measures to prevent, detect, and mitigate such attacks while continuing to provide seamless and efficient service to legitimate callers. However, security measures should not become an impeding factor for establishing end-to-end emergency calls over IP networks.

Issue: Responsibility for the Acquisition and Delivery of Location Information and the Mechanism to Obtain and Verify the Location Data Needs to Be Defined

Impact on NG9-1-1: Automatic location information is essential in an NG9-1-1 environment to route the 9-1-1 call to the appropriate PSAP, dispatch emergency responders to the right location, and allow for future 9-1-1 capabilities. To continue the current level of 9-1-1 service that provides location information automatically and to allow other 9-1-1 features (e.g., accepting 9-1-1 text messages), location information must be obtained and delivered automatically with an NG9-1-1 call for wireline, wireless, VoIP and other IP-based devices, and any future devices.

Stakeholder(s) Affected: Issues related to the acquisition and delivery of location information will affect service providers (such as infrastructure and access service providers), PSAPs, public safety entities, and the Federal Government. All 9-1-1 service providers, including next generation infrastructure and access providers, have responsibility for helping determine and deliver location information. As for PSAPs and first responders, location information is vital in responding to an emergency. Lastly, a national mandate may be necessary to require service providers to obtain and provide accurate location information for VoIP and other IP device calls.

Background and Description: Caller's location information, whether civic or geospatial in nature, is needed to route emergency calls to the most appropriate PSAP, as well as to ensure responder agencies are dispatched to the correct location. With the transition to NG9-1-1, several location data acquisition, conveyance, and determination issues arise, including—

Obtaining and verifying caller location information despite the increased mobility and nomadic natureof the general public

Ensuring that communications devices or systems are capable of delivering location information

Accurately converting between civic and geospatial information

Standardizing the location data structures and types received and provided to the PSAPs.

To date, the responsibility to obtain and verify location information for wireline and wireless 9-1-1 calls has been assigned to the common carriers. However, as in the case of 9-1-1 calls from IP-enabled devices using WiFi or WiMax hotspots, the responsible party has not been identified because such service involves not only a service provider (providing the WiFi service hotspot), but an access provider as well (providing the IP/Internet Service). As such, the responsibility could rest with the service and/or the access provider. Before accurate location information can be obtained, the responsible party must be identified and a mechanism that associates and verifies the location data or reference of the caller needs to be developed and deployed.

Assumptions/Considerations: The analysis of this issue assumes that PSAPs will be able to receive 9-1-1 calls from IP-based devices that ultimately connect to an emergency communication network. There may be a variety of 9-1-1 call originating sources public or private that may connect to NG9-1-1 (e.g., WiFi Hotspot, enterprise local area network). This issue also assumes that E9-1-1 is the baseline of 9-1-1 services available at many PSAPs.

Issue: With the Increase of Personally Identifiable Information Available with a NG9-1-1 Call, Existing Laws May Not Be Sufficient to Protect the Privacy of a 9-1-1 Caller

Impact on NG9-1-1: For public safety to respond to 9-1-1 calls, personal and identifiable information must be obtained both automatically (e.g., location and telephone number) and directly from the caller or from call stream data. The ability to positively identify the location of a caller, coupled with the ability to automatically identify other personal information that may contribute to effective emergency response, but poses privacy concerns that are certainly present today, and will increase in the NG9-1-1 environment. To further compound privacy concerns, NG9-1-1 offers the ability to interconnect other systems or databases to provide supplemental information such as medical and health data that are currently not widely available to PSAPs and public safety agencies in an automated or electronic fashion. This raises issues of legal privacy and the need for NG systems to adequately address user access and data rights management and the ability to respond to compromised access.

Stakeholder(s) Affected: The confidentiality of 9-1-1 call information can be required by state legislation and regulation, and maintained by adequate system controls and mechanisms. State governments are responsible for developing appropriate legislation and regulation allowing or denying the release of 9-1-1 call records (e.g., voice recording, electronic or printed data, or transcriptions) to ensure the caller's privacy. System service and infrastructure providers may be responsible for ensuring that the mechanisms are in place to adequately control user access and manage the rights to data and information.

Background and Description: Data privacy involves safeguarding personally identifiable information (PII) which can be used to uniquely identify or locate a single person. With the implementation of wireless E9-1-1, many privacy advocates raised concerns that the E9-1-1 technology allowed a wireless service provider to determine the location of the caller based on the cellular telephone. Privacy advocates feared that cellular telephones would become tracking devices for the benefit of both the government and private industry.[29] For example, in the NG9-1-1 environment, service providers could implement a solution that positively identifies the location of a 9-1-1 caller using a mobile IP-based device. In addition to tracking location, the NG9-1-1 system will allow digital transmission and storage of a variety of new types of personal information, such as medical and telematics data, photos of victims, and other visual images. The range of PII that can be collected and transmitted may raise concerns from the many interest groups tracking privacy issues because the quantity and quality of information NG9-1-1 will make accessible to the PSAPs and public safety agencies is well beyond what is currently available. Today's patchwork of privacy legislation from federal, state and local governments make it impossible to identify a lowest common denominator for privacy regulations. The challenge will be to develop legislation, proper system controls and capabilities (e.g., who has access), and business processes (e.g., how often is information provided, what information is provided) to minimize privacy concerns while balancing the need for data access against the right for privacy.

Assumptions/Considerations: The analysis of this issue assumes that service providers (e.g., infrastructure and/or access) will automatically obtain and provide location information for callers. In addition, it assumes that some states do not have appropriate privacy laws to protect against the illegal access to and use of a caller's identity and information. Finally, this analysis assumes that external, third-party databases or systems will be accessible to NG9-1-1 to provide supplemental data such as medical information and crash data associated with a 9-1-1 call.

Issue: Routing and Prioritization in the NG9-1-1 Environment Is More Complex with the Ability to Dynamically Route Calls Based on Factors Beyond Location of the Caller

Impact on NG9-1-1: Call routing, a core component of the NG9-1-1 system, is critical to the functionality of the system. With the IP open architecture, the NG9-1-1 System will enable the ability to route 9-1-1 calls throughout the Nation instead of being limited to a specific area as in today's 9-1-1. Location-to-Service Translation (LoST) servers and a national PSAP uniform resource locator (URL) registry must be developed to support the ability to route NG9-1-1 calls at a national level. LoST servers contents the necessary information (e.g., map service identifiers, geospatial or civic location information) to take the caller's location and determine the routing path to the appropriate NG9-1-1 PSAP based on the PSAP's URL. Without appropriate routing, 9-1-1 calls on the NG9-1-1 system may not reach their destination— the PSAPs. If 9-1-1 calls do not reach the PSAPs, it is possible that they may go unanswered.

Stakeholder(s) Affected: Issues related to routing and prioritizing 9-1-1 calls will affect PSAPs, and are beginning to be addressed by Standards Development Organizations (SDOs). Without defined call treatment protocols and a corresponding PSAP URL registry, 9-1-1 calls may be routed to an inappropriate or unavailable PSAP, affecting the ability of the PSAPs to answer 9-1-1 calls in a timely manner, and may not take full advantage of the routing and treatment capabilities of NG9-1-1. It may require PSAPs to re-route or transfer the 9-1-1 call to another PSAP within the jurisdiction that is capable of answering 9-1-1 calls from specific devices. While SDOs are currently in the process of developing standards or protocols that will allow 9-1-1 calls to be prioritized as emergency calls and routed appropriately within call origination networks, additional work is necessary to add other call treatment features including the application of business rules addressing factors like calling devices, PSAP jurisdiction, and others.

Background and Description: PSAPs typically serve a well-defined, but local geographic region and vary in 9-1-1 capabilities ranging from basic (i.e., landline three-digit dialing, circuit-based transmission) to Wireless Phase II 9-1-1 (e.g., receives wireless 9-1-1 calls and automatically identifies caller location). To date, 9-1-1 calls, for the most part, are routed based on a single factor—location of the caller. However, as NG9-1-1 capabilities are phased into the current 9-1-1 capabilities, it is crucial that a call treatment protocol is created to determine how 9-1-1 calls are routed over an IP network to the appropriate PSAP based on the location of the caller, the capability of the PSAP, and business rules. Business rules describe the operational definitions and constraints that can be applied to the system to dynamically specify how the system should react under different circumstances. As an example, if a deaf or hearing impaired caller calls 9-1-1 and potentially requires relay services, business rules must be developed determine how the call will be routed to a PSAP that can automatically conference in an interpreting service.

Assumptions/Considerations: The analysis of this issue assumes that if NG9-1-1 is implemented in a phased approach nationwide, not all PSAPs will have NG9-1-1 capabilities simultaneously. It is also important to note that NG9-1-1 system routing will also depend on establishment of the appropriate regulations, protocols, and databases to support nationwide PSAP call routing and the provision of caller location.

Issue: Mechanisms Need to Be Developed to Certify and Authenticate Service and Infrastructure Providers to Allow Access to NG9-1-1

Impact on NG9-1-1: Telecommunications services and the access to 9-1-1 by such services are potentially much more complex in the NG environment. The actual services themselves, whether they are the delivery of a 9-1-1 call, or an enhancement to the treatment of that call, could be provided by third party service providers separate from those that provide access to such services. Consequently, mechanisms to regulate, certify, and authenticate service providers of all types critical to emergency communications must be defined to ensure consistent, authorized, and standard contributions to the process and NG9-1-1 PSAPs can receive data-related to 9-1-1 calls.

Stakeholder(s) Affected: The regulation, certification, and authentication of all types of service providers that need to gain access to the NG9-1-1 network will affect service providers and the federal and state government. Without mechanisms to connect to the NG9-1-1 network, service providers will not be able to provide the necessary data and service capabilities. Currently no mandate exists that identifies prerequisites that service providers would be required to meet (e.g., security and standards requirements) to be allowed to directly connect to the NG9-1-1 system as the Local Exchange Carriers (LEC) do to the current system.

Background and Description: To date, LECs serving as 9-1-1 service providers are regulated through the Federal Communications Commission (FCC) and individual state regulatory commissions, and are the only entities with direct access to the 9-1-1 functions. To access the 9-1-1 functions, wireless and VoIP providers must access to the Public Switched Telephone Network (PSTN) through such providers. The FCC requires LECs to allow access to the 9-1-1 network through the PSTN to wireless and VoIP providers. However, this method of connecting to the 9-1-1 network can only handle voice 9-1-1 calls. To ensure that only reputable service providers have access to the NG9-1-1 system, certification and authentication mechanisms must be developed to identify and determine which service providers are allowed to access the NG9-1-1 network in a prescribed and standardized manner. The mechanisms may outline service providers' prerequisites, standards, and requirements to protect the security and promote an open architecture of the NG9-1-1 network. Assumptions/Considerations: The analysis of this issue assumes that callers will use IP-based device capabilities such as instant messaging, multimedia messaging services, and photo functions to call 9-1-1 in addition to the traditional voice 9-1-1 call.

Issue: NG9-1-1 Allows for a Level of Coordination and Resource Sharingthat Does Not Currently Exist in Today's 9-1-1 Environment

Impact on NG9-1-1: The nature of NG9-1-1 system allows for the sharing of services and resources, including infrastructure and applications. Without establishing an agreed upon framework (institutional arrangement) among PSAPs, 9-1-1 Authorities, and other entities (e.g., public safety dispatch, service providers, etc.), the benefits of sharing resources and services will not be fully realized. This may prevent the realization of potential cost savings by a region or state that has multiple PSAPs and other entities.

Stakeholder(s) Affected: NG9-1-1 enables interconnection of a variety of services and resources not currently used by PSAPs and 9-1-1 Authorities. Cooperation and collaboration with multiple new partners is necessary to realize the full benefit of NG capabilities. The sharing of service and resources will involve PSAPs, 9-1-1 Authorities, and service providers. The PSAPs and 9-1-1 Authorities will need to establish a framework to manage and coordinate sharing relationships. In addition, service providers may be responsible for managing and maintaining some of the shared resources involved. Lack of coordination has consequences at all levels. Local jurisdictions may be unprepared to upgrade their PSAPs for IP-based communications because of a lack of coordination by state and local stakeholders, both public and private, and may be unable to take full advantage NG9-1-1 features like re-routing and call overflow management. The Nation as a whole may also be unable to gain the benefits of a national interconnected emergency number system without stakeholder coordination at the national, state, regional and local levels including potential economies of scale, and more consistent application of NG9-1-1 across the country.

Background and Description: Currently, many PSAPs are individually managed and upgraded, while in some areas PSAP services and upgrades are coordinated through higher level regional or state authorities. Regardless, the current circuit-based dedicated 9-1-1 system architecture does not promote or support the sharing of resources and services among the PSAPs. However, the NG9-1-1 system is based on an open IP architecture. This allows any PSAP to manage and share resources and services to potentially reduce the cost of upgrading and maintaining NG9-1-1 services. It is important that the PSAPs develop and agree upon terms and conditions that identify what will be shared and who will be responsible for operation and maintenance issues. In addition to establishing institutional arrangements among PSAPs, service providers should be involved because they may be responsible for managing and maintaining infrastructure or databases that will be used by the PSAPs. For example, components of the NG9-1-1 system for which responsibility may be shared between public (e.g., PSAPs) and private (e.g., service providers) entities include the PSAP URL registry and routing databases (e.g., LoST). The service providers may be responsible for managing and maintaining the databases, while the PSAPs may be responsible for updating the database based on their jurisdictional boundaries and capabilities. These opportunities for sharing need to be identified and codified to determine who will be responsible for the resource and how the resource will be managed, maintained, and staffed.

Assumptions/Considerations: The analysis of this issue assumes that in an NG9-1-1 environment, the roles of the PSAPs, responders, and related entities are expected to expand beyond traditional 9-1-1 services with higher levels of interaction, managed situational intelligence, enhanced capabilities, and more comprehensive communication and coordinated response services.

Issue: Industry May Be Reluctant to Develop and Adopt Open Standards which Could Limit the Availability of Interoperable NG9-1-1 Services and Equipment

Impact on NG9-1-1: Lack of support from service providers and equipment manufacturers in developing and adopting open standards will limit the availability of interoperable services and equipment. This will have an impact on the ability of the NG9-1-1 system to interconnect and achieve the NG9-1-1 vision to provide ubiquitous, interconnected NG9-1-1 services across the Nation.

Stakeholder(s) Affected: The availability of open, nonproprietary standards and technology will affect PSAPs and 9-1-1 Authorities procuring equipment and services to upgrade to NG9-1-1. For entities choosing to take an independent approach to NG9-1-1 implementation, the risk of implementing proprietary solutions that do not interoperate with other 9-1-1 and NG9-1-1 systems increases. Some PSAPs will face difficult decisions about which technology to embrace when there are competing proprietary platforms to choose from and not fully realize the ramifications of that decision until a later time when their network cannot be integrated into NG9-1-1.

Background and Description: Open architecture allows for the ability of system architects and maintainers to incorporate new components, install upgrades, or swap one manufacturer's device for another's without being restricted by proprietary constraints. However, given the relatively small market for 9-1-1 equipment compared with business and consumer electronics, and/or the lack of timely and effective open standards, service and equipment providers may choose to move forward by offering proprietary solutions to prospective buyers. Furthermore, to upgrade to new open architecture hardware and software for a small market such as 9-1-1 may make it difficult for service and equipment providers to meet profit margin goals because the cost associated with upgrading may out weight the revenues realized. Similarly, first responder agencies are facing challenges in getting manufacturers to adopt standards for private land mobile radio systems in order to promote interoperability among federal, state, and local emergency responders. Lack of support from service providers and equipment manufacturers in adopting open standards will limit the availability of interoperable services and equipment.

Assumptions/Considerations: The analysis of this issue assumes that the NG9-1-1 system architecture is based on open, non-proprietary standards and technology. In addition, it assumes that implementers of NG9-1-1 solutions will use open architecture to promote interoperability among disparate systems.

Issue: Liability Protection or Parity Needs to Be Extended to Protect Stakeholders Involved in NG9-1-1

Impact on NG9-1-1: Experience in the deployment of E9-1-1 has shown that a lack of legal clarity on the issue of liability parity can lead to delays in the provisioning of E9-1-1 service. NG9-1-1 will potentially promote a more complex service delivery environment, which will further complicate how liability protection is appropriately provided for new and future services.

Stakeholder(s) Affected: In the current system, 9-1-1 wireless and wireline service providers are afforded protection from liability by legislative action. With the transition to NG9-1-1, new types of service providers will be entering the system. Their standing with respect to liability has not been defined by law. This may prevent the entry of new parties and the provision of data that could speed the delivery of more effective emergency response services.

Background and Description: Within NG9-1-1, a call may include multimedia information containing essential information about the emergency situation that may be necessary to appropriately route the call. The NG9-1-1 call stream could also contain supportive or supplemental data (e.g., weather, traffic information, and medical data) obtained through free or paid services. These third-party sources traditionally make their information available to the public or to a limited set of subscribers with access to sensitive information, such as PSAPs. Many of the third-party information sources from which the services will be built are only now being developed. The providers of these data should be aware of the potential for use of their services by PSAP call takers, public safety dispatchers, and first responders. Examples of such information include map displays and other integrated geospatial data, such as those that are becoming available that integrate maps, three-dimensional images, and detailed information about a particular location of the caller. Another example is vehicle telematics crash data that could be used to make dispatch decisions about the quantity and type of responders, and the triage of crash victims to the appropriate medical facility.

Assumptions/Considerations: The analysis of this issue assumes that some 9-1-1 Authorities will be reluctant, or even refuse, to complete VoIP emergency calls because they lack the legal safeguards that protect them from liability which exist today for wireline and wireless emergency calls.

Issue: New Capabilities and Services Enabled by NG9-1-1 Could Impact PSAP Operations and Training for Both PSAP Administrators and Call Takers

Impact on NG9-1-1: As implementation of the NG9-1-1 occurs, 9-1-1 Authorities and PSAP directors will have to determine how to modify existing PSAP practices and procedures (e.g., standard operating procedures [SOP] and training) and resources (e.g., human machine interface [HMI] solutions for existing customer premises equipment [CPE]) to accommodate NG9-1-1 services. Currently, neither nationwide SOPs nor a training curriculum fully exist NG9-1-1 system operation.

Stakeholder(s) Affected: It is anticipated that the demands on PSAP call takers are likely to increase with the complexity of call taking functions caused by new types of multimedia data. Likewise, 9-1-1 Authorities and PSAPs directors will face the challenge of configuring systems and training call takers to accept and process this new NG9-1-1 data.

Background and Description: As a result of the introduction of the NG9-1-1, PSAP operations will face new challenges associated with the increase in real-time multimedia information—text, still images, or video in addition to voice—passing between a person needing assistance and the call taker. While NG9-1-1 will provide new and enhanced tools to deal with such complexity, the increased quantity of multimedia data will complicate existing call taking functions. Receipt of calls from IP-based communication devices, in addition to conventional wireline calls, could test call takers' ability to maintain their quick and accurate response to emergency situations. In addition, the NG9-1-1 environment will influence how 9-1-1 Authorities and PSAP directors support PSAP operations (e.g., including configuring business rules and systems to limit transmission of extraneous data transactions that can slow response times). Within the NG9-1-1 system, once a call is answered, the PSAP call takers may face increased stress and workloads because of the wide range of information types that could be delivered in each 9-1-1 call.[30] Moreover, the increase in multimedia data may make call records management[31] more challenging because NG9-1-1 will present new information types and formats to archive, such as audio, video, text, still imagery, and other data types.[32]

Assumptions/Considerations: The analysis of this issue assumes that responsibilities of 9-1-1 Authorities will likely expand, particularly with regard to configuring the NG9-1-1 system for their PSAP. While the PSAP director's responsibilities will also likely increase, it is assumed that they will be more focused on the daily operations of their specific PSAP and training of call takers to ensure that their PSAP meets the requirements set forth by their 9-1-1 Authority. It also assumes that even though it is anticipated that the NG9-1-1 system will include HMI solutions to assist call takers in answering and processing call data, they may still need to analyze and make decisions regarding what information should be transferred to the dispatchers and first responder agencies based on their training and experience.

CONCLUSION

The USDOT has identified potential transition issues to be addressed at the national, state, regional, and local levels before and during the implementation of NG9-1-1. As noted earlier, the level of impact and priority of each issue depends on the implementation pathway employed by the state, regional, or local 9-1-1 Authorities. These deployment scenarios will be further evaluated in the NG9-1-1 Transition Plan to identify strategies that can be adapted and used by the 9-1-1 community in implementing NG9-1-1 across the country. Collectively, the analysis of the issues and the NG9-1-1 Transition Plan will provide a foundation for local and state 9-1-1 Authorities in planning and deploying NG9-1-1.

APPENDIX A—DEFINITION OF NG9-1-1 STAKEHOLDERS

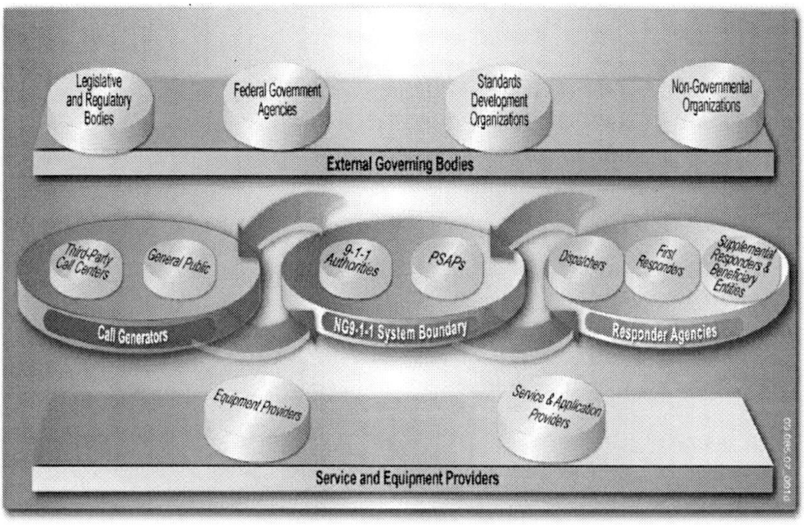

Stakeholder roles may change in the transition from today's 9-1-1 environment to the NG9-1-1 environment; however, the extent of change will vary among the types of stakeholders. In this context, "type of stakeholder" refers to an entire stakeholder category rather than individual entities within that category (e.g., responder agencies versus individual first responders). To better understand the fundamental roles and functions of each 9-1-1 stakeholder, the following subsections provide a high-level description of each stakeholder type.

Service and Equipment Providers

Service and equipment providers are responsible for developing, supplying, and sometimes maintaining the hardware, software, and other equipment used by PSAPs in transferring 9-1-1 calls; providing public access to 9-1-1 services, physical and network access, and telecommunications and IT equipment; and helping to maintain the data and information necessary to route and deliver emergency calls. Types of service and equipment providers include 9-1-1 service providers, service access providers, service application providers, and equipment suppliers.

Third-Party Call Center

Third-party call centers receive voice, video, text, images, and other data originating from the public via various services, such as automobile telematics, relay services, medical alert, poison control hotlines, and satellite telecommunications services. These call centers intercede between the general public and 9-1-1 services for a growing number of emergency calls to help support the delivery and processing of the calls involved.

9-1-1 Authority

A 9-1-1 authority has jurisdiction over and/or supports a particular 9-1-1 system. The 9-1-1 authority could be a county/parish or city government, a special 9-1-1 or Emergency Communications district, a Council of Governments, an individual PSAP, a statewide agency, or other similar body. The 9-1-1 authority generally manages human resource requirements and activities, oversees service delivery, performs high-level database management and support functions, oversees funding and procurement of supporting infrastructure and services, establishes SOPs and operational policies, and is responsible for security at the PSAPs.

Public Safety Answering Point

A PSAP is the generic name for an emergency communications center agency that receives, processes, and transfers 9-1-1 calls. Some PSAPs also direct or dispatch 9-1-1 or other emergency calls to appropriate police, fire, and emergency medical services (EMS). PSAPs throughout the Nation have different levels of capabilities and functions and are typically categorized as follows: "Greenfield" 9-1-1, 33 Basic 9-1-1, E9-1-1, or Wireless E9-1-1.[33]

PSAP Call Taker

A PSAP call taker is a person who receives emergency and non-emergency calls, determines the nature of caller situations, elicits the location of the emergency and other necessary information, and relays essential information to dispatchers, staff, and other agencies as needed, using telephony and computer equipment. In some instances, the call taker may also be the dispatcher, and in that case, the call taker/dispatcher communicates directly with the responders. In other cases, the call taker may provide emergency instructions to the caller prior to the arrival of first responders.

Responder Agencies

Responder agencies—dispatch entities, law enforcement, fire and rescue, EMS, supplemental responders, and beneficiary entities—have a legal or consensual obligation to respond to or obtain information from emergency calls to support comprehensive incident management.

Legislative and Regulatory Bodies

Legislative bodies exist at the federal, state, and local levels to draft and pass legislation for the welfare and benefit of the public. Of interest to this report is legislation related to the 9-1-1 services. To ensure that the provisions of the legislation are implemented and enforced, regulatory bodies at each level of government may write draft regulations that are open to public comment, but binding on the public once promulgated in final form. The regulatory body itself or a court having the appropriate jurisdiction may impose sanctions for regulatory infractions. For example, revising tariffs, mandating a nationwide IP-enabled emergency network, enforcing IP security standards to ensure privacy of information transmitted on the NG9-1-1 System, and regulating how 9-1-1 fees are collected can be completed in a way that is consonant with the entire policy context for 9-1-1. Examples of legislative and regulatory bodies include state legislatures, Public Utility Commissions (PUCs), and the FCC.

Federal Government Agencies

Federal Government agencies are responsible for establishing policies and funding, providing leadership, and promoting coordination and communications between agencies and organizations involved with 9-1-1 services. The Federal Government also provides a framework for state and local governments to transition from the current 9-1-1 system to the NG9-1-1 System. The most prominent government organizations involved in 9-1-1 include the USDOT

Intelligent Transportation Systems Joint Program Office, Department of Homeland Security, USDOT National Highway Traffic Safety Administration–National Telecommunications and Information Administration National 9-1-1 Implementation and Coordination Office, Department of Justice, FCC, and Department of Health and Human Services.

Non-Governmental Organizations

Non-governmental organizations include those agencies and organizations that are considered part of the private sector. Non-governmental organizations play a critical role in the implementation of the NG9-1-1 System because of their involvement in identifying the needs of their membership, thereby shaping the products and services used by consumers and PSAPs who use the NG9-1-1 System. Examples of non-governmental organizations include SDOs, professional and industry associations, citizen and special interest advocacy organizations, private emergency response and recovery organizations, and research and academia.

General Public

The general public relies heavily on the ability to access 9-1-1 service with full functionality through any communications device during an emergency. The public expects a timely response from the emergency responder agencies once a call is made to 9-1-1 and determines the success of the 9-1-1 service based on the response time. To date, the general public can make 9-1-1 calls through wireline telephone, wireless telephone, teletype/telecommunications device for the deaf (TTY/TDD) over wireline, and VoIP, and indirectly through third-party call centers associated with telematics and relay services.

Appendix B—Acronym List

ACD	Automatic Call Distribution
CPE	Customer Premises Equipment
E9-1-1	Enhanced 9-1-1
EMS	Emergency Medical Services
ESNet	Emergency Services Network
FCC	Federal Communications Commission

HMI	Human Machine Interface
IP	Internet Protocol
LEC	Local Exchange Carrier
LIS	Location Information Server
LoST	Location-to-Service Translation
NG9-1-1	Next Generation 9-1-1
PII	Personally Identifiable Information
PSAP	Public Service Answering Point
PSTN	Public Switched Telephone Network
PUC	Public Utility Commission
SDO	Standards Development Organization
SOP	Standard Operating Procedure
TTY/TDD	Teletype/Telecommunications Device for the Deaf
USDOT	U.S. Department of Transportation
VoIP	Voice over Internet Protocol
WiFi	Wireless Fidelity

Appendix B: Acronyms

AIP	Access Infrastructure Provider
ALI	Automatic Location Identification
ANI	Automatic Number Identification
APCO	Association of Public-Safety Communications Officials
ASL	American Sign Language
CLEC	Competitive Local Exchange Carrier
CPE	Customer Premises Equipment
CONOPS	Concept of Operations
DHS	Department of Homeland Security
DoJ	Department of Justice
DoS	Denial of Service
E9-1-1	Enhanced 9-1-1
ECRIT	Emergency Context Resolution with Internet Technologies
EDXL	Emergency Data Exchange Language
FCC	Federal Communication Commission
HHS	Department of Health and Human Services
HMI	Human Machine Interface
ICO	Implementation and Coordination Office
IETF	Internet Engineering Task Force

IP	Internet Protocol
IT	Information Technology
ITS	Intelligent Transportation Systems
JPO	Joint Programs Office
LEC	Local Exchange Carrier
LoST	Location-to-Service Translation Protocol
MOU	Memorandum of Understanding
NENA	National Emergency Number Association
NG9-1-1	Next Generation 9-1-1
NHTSA	National Highway Traffic Safety Administration
NRIC	National Reliability and Interoperability Council
NTIA	National Telecommunications and Information Administration
ODC	Operational Development Committee
PII	Personally Identifiable Information
PSAP	Public Safety Answering Point
PSTN	Public Switched Telephone Network
RUS	Rural Utilities Service
SDO	Standards Development Organizations
SIP	Session Initiation Protocol
SOP	Standard Operating Procedure
TDD/TTY	Telecommunications Device for the Deaf / Teletypewriter
USDOT	U.S. Department of Transportation
VoIP	Voice over Internet Protocol
VRI	Video Remote Interpreting
XML	Extensive Markup Language

APPENDIX C: GLOSSARY

• 9-1-1	• A three-digit telephone number to facilitate the reporting of an emergency requiring response by a public safety agency.
• 9-1-1 System	• The set of network, database, and customer premises equipment (CPE) components required to provide 9-1-1 service.
• Activity	• See "Functional Activity."
• Analog	• Continuous and variable electrical waves that represent an infinite number of values; as opposed digital.
• Association for Public-Safety Communications—International (APCO)	• A not-for-profit organization established in 1935 and the world's largest organization dedicated to public safety communications. Members rely on APCO for their professional needs—from examining standards and issues to providing education, products and services, and frequency coordination

	services.
• Authentication	• Determination or verification of a user's identity and/or the user's eligibility to access to a system, network, or data; measures to prevent unauthorized access to information and resources.
• Automatic Call Distributor (ACD)	• Equipment or application that automatically distributes incoming calls to available PSAP call takers in the order the calls are received, or queues calls until a call taker becomes available.
• Automatic Collision Notification (ACN)	• The process of identifying that a motor vehicle has been involved in a collision, collecting data from sensors in the vehicle, and communicating that data to a PSAP.
• Automatic Event Alert	• 9-1-1 calls placed by sensors or similar initiating devise. Includes alarms, telematics, and sensor data, and may also include real-time communications.
• Automatic Location	• The automatic display at the PSAP of the caller's telephone number, the address or
• Identification (ALI)	• location of the telephone, and supplementary emergency services information.
• Automatic Location Identification (ALI) Database	• The set of ALI records residing on a computer system.
• Automatic Number Identification (ANI)	• Telephone number associated with the access line from which a call originates.
• Availability	• The operational ability of necessary and beneficial data interfaces to support call processing and emergency response; or the amount or percentage of time that the system provides service.
• Backup Public Safety Access Point (Backup PSAP)	• Typically, a disaster recovery answering point that serves as a backup to the primary PSAP and is not collocated with the primary PSAP.
• Border Control Function (BCF)	• BCF activities create a boundary between the internal network resources and the external network(s). Access to particular network resources behind a BCF-enabled gateway, can be restricted by a variety of methods. Most BCFs offer a level of Network Address Translation (NAT) and provide firewall-like functions. The deployment of BCFs at the edge of the network can secure and protect the system from outside resources by creating a Demilitarized Zone (DMZ) that protects the internal network resources from the outside network. The DMZ allows access only to the trusted parties that authenticate to the network. BCFs can also offer network-to-network interface functions for allowing traffic to be delivered across the network and Session Initiation Protocol (SIP) session border control functionality.
• Business Rules	• Business rules describe the operational definitions and constraints that can be applied to the system that dynamically specify how the system should react under different circumstances. Business rules are used throughout the NG9-1-1 System enabling jurisdictions and 9-1-1 authorities to configure the system based upon the needs of the locality or region and can be modified or updated as needed.

• Call	• For the purposes of this NG9-1-1 report, any real-time communication—voice, text, or video—between a person needing assistance and a PSAP call taker. This term also includes non-human-initiated automatic event alerts, such as alarms, telematics, or sensor data, which may also include real-time communications.
• 9-1-1	• A three-digit telephone number to facilitate the reporting of an emergency requiring response by a public safety agency.
• Callback	• The ability to re-contact the calling party.
• Call Delivery	• The capability to route a 9-1-1 call to the designated selective router for ultimate delivery to the designated PSAP for the caller's ANI/KEY.
• Call Detail Record	• All system (including network) data accessible with the delivery of the call, and all data automatically added as part of call processing. This includes Essential Data (including reference key to network component and call progress records) and Supportive Data. Part of the Call Record.
• Caller Location Information	• Data pertaining to the geospatial location of the caller, regardless of whether the caller is a person or an automatic event alert system.
• Call Narrative	• Supplemental Data (or caller-generated data) manually gathered and entered by the call taker for the purposes of documenting the call. Part of the Call Record.
• Call Record	• The collection of all information related to a call (including Essential, Supportive, and Supplemental data); composed of Call Detail Record, Call Recording, and Call Narrative.
• Call Recording	• The electronic documentation of the interactive communication (e.g., audio, video, text, image) between the caller, call taker, and any conferenced parties. Part of the Call Record.
• Call Routing	• The capability to selectively direct the 9-1-1 call to the appropriate PSAP.
• Call Taker	• As used in 9-1-1, a person (sometimes referred to as a telecommunicator) who receives emergency and non-emergency calls by telephone and other sources, determines situations, elicits necessary information, and relays essential information to dispatches, staff, and other agencies, as needed, using telephony and computer equipment.
• Call Transfer	• The capability to redirect a call to another party.
• Call Type	• Classification of a 9-1-1 call that indicates the call access method, which can affect call treatment, routing, and processing. Call types may include voice caller, short message service (SMS) text, Simple Mail Transfer Protocol (SMTP) text, multimedia, telematics data, ANI, silent alarms, etc.
• Circuit-Switch	• The establishment, by dialing, of a temporary physical path between points. The path is terminated when either end of the connection sends a disconnect signal by hanging up.
• Civic Address Information	• Street address data, inclusive of suite/office number, where appropriate.
• Commercial Mobile Radio Service (CMRS)	• A category of wireless telephone service regulated by the FCC (47 CFR 20.9). It includes both cellular and Personal Communications Services (PCS) telephone service.
• Configurability	• Property of a system that supports the rearrangement of interfaces and functionalities.

• Continuity of Operations (COOP)	• A system's ability to prevent critical system failures (e.g., via component redundancy) and to seamlessly conduct updates and repairs.
• Cross-System Authentication	• Authentication across a number of systems or networks via a single authentication process, sometimes referred to as Single Sign-On (SSO), and potentially achieved via proxy authentication.
• Customer Premises Equipment (CPE)	• Communications or terminal equipment located in the customer's facilities; terminal equipment at a PSAP.
• Database	• An organized collection of information, typically stored in computer systems, composed of fields, records (data), and indexes. In 9-1-1, such databases include the master street address guide, telephone number, and telephone customer records.
• Data Integrity	• The property of not having been altered or destroyed in an unauthorized manner.
• Digital	• Relating to calculation, storage, or transmission by numerical methods or discrete units, as opposed to the continuously variable analog. Computerized.
• Disaster Dispatcher	• Any event that can cause a significant disruption to normal emergency calling capability.
	• As used in public safety, a person responsible for receiving and transmitting information pertaining to requests for emergency service and other related activities, tracking vehicles and equipment, and recording other important information using a telephone, radio, and other communications resources.
• 9-1-1	• A three-digit telephone number to facilitate the reporting of an emergency requiring response by a public safety agency.
• Dispatch Operations	• The distribution of emergency information to responder organizations responsible for delivery of emergency services to the public.
• Emergency Call	• A telephone request for public safety agency emergency services that requires immediate action to save a life, to report a fire, or to stop a crime. May include other situations as determined locally.
• Emergency Location	•
• Information	• Data pertaining to the location of the emergency, which may be different from the caller location.
• Emergency Medical Service (EMS)	• A system providing pre-hospital emergency care and transportation to victims of sudden illness or injury.
• Emergency Response	• An effort by public safety personnel and citizens to mitigate the impact of an incident on human life and property.
• Enhanced 9-1-1 (E9-1-1)	• An emergency telephone system that includes network switching, database, and CPE elements capable of providing selective routing, selective transfer, fixed transfer, caller routing and location information, and ALI.
• Enterprise	• The highest level of system functionality.
• Essential Call Data	• Data that support call delivery and adequate response capability. These data, or a reference to them, is automatically provided as a part of call or message initiation. Examples include location, callback data, and call type.

• Extensibility	• The property of a system to be adaptable for future growth. The ability to add extended functionality to a system.
• Fixed Transfer	• The capability of a PSAP call taker to direct a 9-1-1 call to a predetermined location by depressing a single button.
• Firewall	• The primary method for keeping a computer secure from intruders. It allows or blocks traffic into and out of a private network or the user's computer.
• Functional Activity	• Bounded piece of work to be performed that describes the people, processes, and technology used.
• Gateway	• The point at which a circuit-switched call is encoded and repackaged into IP packets; equipment that provides interconnection between two networks with different communications protocols; two examples are packet assembler/disassemblers and protocol converters.
• Geographic Information System (GIS)	• A computer software system that enables one to visualize geographic aspects of a body of data. It contains the ability to translate implicit geographic data (such as a street address) into an explicit map location. It has the ability to query and analyze data in order to receive the results in the form of a map. It also can be used to graphically display coordinates on a map (i.e., latitude/longitude) from a wireless 9-1-1 call.
• Global Positioning System (GPS)	• A satellite-based location determination technology.
• Human Machine Interface (HMI)	• Graphical, textual, and auditory means by which an end user interacts with a system. HMI within 9-1-1 systems enables direct interaction between the end-user (human) and a system (computer, machine). For the call takers at the PSAP, HMI provides them with the ability to manipulate the system via commands and inputs, and receive an output from the system based on specified criteria.
• Integrity	• See "Data Integrity."
• International Telecommunications Union (ITU)	• The telecommunications agency of the United Nations established to provide worldwide standard communications practices and procedures. Formerly CCITT.
• Internet Engineering Task Force (IETF)	• The lead standards-setting authority for Internet protocols.
• Internet Protocol (IP)	• The set of rules by which data are sent from one computer to another on the Internet or other networks.
• Interoperability	• The capability for disparate systems to work together.
• 9-1-1	• A three-digit telephone number to facilitate the reporting of an emergency requiring response by a public safety agency.
• Landline	• Colloquial term for the Public Switched Telephone Network access via an actual copper or fiber optic transmission line that located underground or on telephone poles. Used to differentiate the "wireless" connectivity of a cellular or personal communications services system. Also referred to as "wireline."
• Local Exchange Carrier (LEC)	• A telecommunications carrier under the state/local Public Utilities Act that provides local exchange telecommunications services. Also known as Incumbent Local Exchange Carrier (ILEC), Alternate Local Exchange Carrier (ALEC), Competitive Local Exchange Carrier (CLEC),

	Competitive Access Provider (CAP), Certified Local Exchange Carrier (CLEC), and Local Service Provider (LSP).
• Location	• See "Caller Location Information" and "Emergency Location Information."
• National Emergency Number Association (NENA)	• A not-for-profit corporation established in 1982 to further the goal of "One Nation– One Number." NENA is a networking source and promotes research, planning, and training. It strives to educate, set standards, and provide certification programs, legislative representation, and technical assistance for implementing and managing 9-1-1 systems.
• Nature of Emergency Network	• Reason for a citizen's request for response from emergency services (e.g., heart attack, vehicle collision, burglary) An arrangement of devices that can communicate with each other.
• Overflow	• The telecommunications term for the condition when there are more calls than the primary network path is designated to handle. This condition invokes the need to perform some form of call treatment, such as busy signals or alternate routing.
• Packet	• Logical grouping of information that includes a header containing control information and (usually) user data. Packets are most often used to refer to network layer units of data. The terms *datagram*, *frame*, *message*, and *segment* are also used to describe logical information groupings at various layers of the Operating System Interface (OSI) reference model and in various technology circles.
• Packet-Switch	• A network technology that breaks up a message into small packets for transmission. Each packet contains a destination address. Thus, not all packets in a single message must travel the same path. As traffic conditions change, they can be dynamically routed via different paths in the network, and they can even arrive out of order. The destination computer reassembles the packets into their proper sequence.
• Personal Digital Assistant (PDA)	• Small, handheld device used to store address book information, telephone numbers, personal contacts, and other personal information.
• Protocol	• A set of rules or conventions that govern the format and relative timing of data in a communications network. There are three basic types of protocols: characteroriented, byte-oriented, and bit-oriented. The protocols for data communications cover such activities as framing, error handling, transparency, and line control.
• Public Safety Answering Point (PSAP)	• A facility equipped and staffed to receive 9-1-1 calls; a generic name for a municipal or county emergency communications center dispatch agency that directs 9-1-1 or other emergency calls to appropriate police, fire, and emergency medical services agencies and personnel
• Public Switched Telephone Network (PSTN)	• The network of equipment, lines, and controls assembled to establish communication paths between calling and called parties in North America.
• PSTN UA	• Typically a traditional telephone, but can also be a TDD/TTY (Telecommunications Device for the Deaf or Teletype device).
• Redundancy	• Duplication of components, running in parallel, to increase reliability; a backup system (either a device or a connection) that serves in the event of primary system failure.

• Reliability	• The ability of a system or component to perform its required functions under stated conditions for a specified period of time.
• Requirement	• A statement of a characteristic that the system must possess in order to be acceptable; the desired system is defined as one that fulfills all of the requirements.
• Router	• An interface device between two networks that selects the best path to complete the call even if there are several networks between the originating network and the destination.
• 9-1-1	• A three-digit telephone number to facilitate the reporting of an emergency requiring response by a public safety agency.
• Scalability	• The property of a system to be readily enlarged, e.g., by adding hardware to increase capacity or throughput.
• Security	• The ability to provide adequate data and service protection to mitigate unauthorized access, service exploitation, and leakage of confidential or sensitive information.
• Selective Routing	• Direction of a 9-1-1 call to the proper PSAP based on the location of the caller.
• Selective Transfer	• The capability to convey a 9-1-1 call to a response agency by operation of one of several buttons typically designated as police, fire, and emergency medical.
• Service Provider	• An entity providing one or more of the following 9-1-1 elements: network, CPE, or database service.
• Session Initiation Protocol (SIP)	• A signaling protocol used to exchange data (including voice, video, text) among an association of participants. [RFC 3261]
• Short Message Service (SMS)	• A text message service that enables messages generally no more than 140–160 characters in length to be sent and transmitted from a cellular telephone. Short messages are stored and forwarded at SMS centers, allowing their retrieval later if the user is not immediately available to receive them.
• Spatial	• Concept of describing a space or area of space.
• Stakeholder	• An individual or group with an interest in the successful delivery of intended results by a project.
• Supplemental Call Data	• Information that may complement, but is not necessary for, call handling and dispatch. This data typically would be automatically or manually queried after the call is delivered to the call taker. Examples include contact information for someone who should be notified of a medical emergency, building blueprints, other addresses in the immediate vicinity, etc.
• Supportive Call Data	• Information beyond essential data that may support call handling and dispatch. The addition of this data to the call stream is triggered by one or more of the data or reference items in essential data for a given call type. An example is ACN data such as "vehicle rollover."
• System of Systems	• Interconnected and decentralized system of interoperable networks.
• Telecommunications Industry Association (TIA)	• A lobbying and trade association, which is the result of the merger of the USTA (United States Telephone Association) and the EIA (Electronic Industries Association).

• TCP (Transmission Control Protocol)	• The set of rules within the TCP/IP protocol suite that ensures that all data arrives accurately and 100-percent intact at the destination.
• Telematics	• The system of components that supports two-way communications with a motor vehicle for the collection or transmission of information and commands.
• Telephony	• The electronic transmission of the human voice.
• Transfer	• A feature that allows PSAP call takers to redirect a 9-1-1 call to another location.
• Transmission Control Protocol/Internet Protocol (TCP/IP)	• A layered set of protocols (sets of rules) used to connect dissimilar computers together. TCP provides the transport service required by the application layer. The TCP layers in the two host computers that are sending data will communicate with each other to ensure reliable data packet transport. IP provides the service user to deliver the datagram to its destination, providing the routing through the network and the error messages if the datagram is undeliverable.
• User Authentication	• See "Authentication."
• Voice over Internet Protocol (VoIP)	• A set of rules that provides distinct packetized voice information in digital format using the Internet Protocol. The IP address assigned to the user's telephone number may be static or dynamic.
• Wireless	• In the telecommunications industry, typically refers to mobile telephony and communications through handheld devices that make a connection using radio frequency (in particular frequency bands often reserved for mobile communications) for personal telecommunications over long distances.
• Wireline	• Standard telephone and data communications systems that use in-ground and telephone pole cables. Also known as landline or land-based.

APPENDIX D: STRATEGIC OPTIONS CATEGORIZED BY STAKEHOLDERS

• Federal Government	
• Funding	• Ensure NG9-1-1 upgrades are considered a fiscal priority for states and local jurisdictions and the Federal Government
	• Encourage state governments and legislatures to give fiscal priority to NG9-1-1 upgrades (based on nationally accepted standards and coherent statewide plans).
	• Encourage states to designate a state agency or other effective state-level mechanism to be responsible for coordinating statewide 9-1-1, where such a mechanism does not already exist.
	• Consider requiring states to include 9-1-1 in their radio interoperability plans and to provide for shared emergency services internetworks.

	• Transform current 9-1-1 funding mechanisms to resolve— • Diminishing revenue base • Disparities with cost recovery • Funding allocation models for shared resources
	• Help States by providing grant subsidies for the capital costs of NG9-1-1 planning, design, procurement, and implementation.
	• Consider legislation that allows use of 9-1-1 fees to pay for NG9-1-1 portion and use of an underlying IP-based emergency service internetwork.
	• Ensure 9-1-1 funds are preserved for 9-1-1 and emergency communication systems
	• Encourage state and local 9-1-1 Authorities to use 9-1-1 funds, surcharges, and fees for costs solely attributable to 9-1-1 operations, services, and equipment (or their statutory purposes).
	• Clearly define legislatively what constitutes 9-1-1 related communications services and the proper use of 9-1-1 fee revenue.
	• Consider tying eligibility for 9-1-1 grant funding to states that do not practice diversion of 9-1-1 funds.
	• Consider mandating that 9-1-1 funds be used exclusively for the provision of 9-1-1 emergency communications services (e.g., 9-1-1, E9-1-1, NG9-1-1, and related or future upgrades).
• Operations	• Prepare and train call takers to handle increased quantity and quality of information available with a NG9-1-1 call
	• Consider conducting a nationwide annual survey to obtain input from call takers and their immediate supervisors on the effectiveness of training, SOPs, and system operations to ensure the quality of service for handling 9-1-1 calls is maintained.
	• Establish funding methods or a grant program to provide necessary financial support so users can be properly trained and outfitted with the necessary training materials.
	• Develop a public awareness program promoting public safety communications as a rewarding career opportunity in an effort to improve PSAP staffing levels.
	• Prepare 9-1-1 Authorities to handle NG9-1-1 system administration, including configuration management, database management, quality assurance, and SOPs
	• Develop implementation, and operations and maintenance best practices, standards, and lessons learned and share them among 9-1-1 Authorities.
	• Develop model strategic plans to help 9-1-1 Authorities at all levels manage NG9-1-1 migration (e.g., interagency coordination, training, security).
	• Prepare 9-1-1 Authorities and PSAP Administrators to handle contingency planning and use of "virtual PSAPs"
	• Develop PSAP operations best practices and lessons learned and share them among 9-1-1 Authorities and PSAPs.
• Standards and Technology	• Complete and accept NG9-1-1 open standards and understand future technology trends to encourage system interoperability and emergency data sharing

	• Coordinate with 9-1-1 stakeholders (e.g., SDOs, private and public stakeholders) to identify all standards work and technology development currently underway within relevant communications.
	• Coordinate with SDOs to conduct a gap analysis of existing standards development work against the industry-accepted architectural model and identify standards that will need to be developed and/or are already developed standards that are inconsistent with the model.
	• Identify and encourage appropriate SDOs and industry groups to address standards work that must be completed or modified to ensure a comprehensive standards and technology environment for NG9-1-1.
	• Identify and leadership/coordinating entity and apply funding to support dedicated attention to the development of standards and technologies considered essential to NG9-1-1 and coordinate emergency data standard coordination and harmonization.
	• Encourage states to enact laws or regulations that mandate the use of open standards for NG9-1-1 systems.
	• Assist in the development of a certification program to ensure that NG9-1-1 services and equipment meet the open standards defined by the industry-accepted architectural model.
	• Establish system access and security controls to protect and manage access to the NG9-1-1 system of systems
	• Consider initiating establishment of new security regulations or modification of existing security regulations to promote consistency among states.
	• Identify and modify, as necessary, federal legislative or regulatory provisions that may constrain the transition to the NG9-1-1 environment, thereby enabling unbundled service offerings.
	• Adopt a national-level certification and authentication process determined by the SDOs to ensure service providers and 9-1-1 Authorities meet the security and system access requirements.
	• Identify and leverage industry and defense and homeland security solutions, standards, and best practices to ensure a consistent and appropriate level of security throughout the interconnected environment of NG9-1-1.
	• Promote the development of identity and data rights access and management to control access to the NG9-1-1 system and data.
	• Determine responsible entity and mechanisms for location determination and acquisition
	• Develop a national mandate that requires service providers (access and transport network and devices) to meet specific standards or requirements for obtaining and providing accurate location information for all call types that access NG9-1-1.
	• Determine the responsibility for location validation and management of national and/or regional databases necessary for NG9-1-1.
	• Examine issue of who within the service provider community will be responsible for providing accurate location information for NG9-1-1 calls.

• Government • and Policy	• Clarify jurisdictional frameworks, responsibilities, and coordination required at all levels of government to enable NG9-1-1
	• Identify a lead agency or central sponsor for guidance and coordination of NG9-1-1 at the national level.
	• Strongly encourage states to convene their state 9-1-1 Administrator, state homeland security/emergency management director, and state chief information officer (CIO) to plan for establishment of NG9-1-1 and its underlying emergency services inter-networks
	• Consider modifying current legislation to assign state/regional 9-1-1 Authorities the responsibility and authority to transition to and manage the NG9-1-1 system.
	• Facilitate appropriate regional or national working groups and orums at all levels to coordinate business rules and data rights management with emergency systems at the state and local levels.
• Government and Policy	• Update regulations, legislation, and other policies to— eflect communications capabilities in an NG9-1-1 system rotect the confidentiality of new types of PII • Extend liability protection and parity for PSAPs and new services
	• Update federal laws and encourage the modification of state laws that may prohibit the use of location information for call routing beyond those covered under traditional 9-1-1 (e.g., include appropriate N-1-1 entities and 800-type emergency provisioning, such as poison control centers and suicide hotlines)
	• Develop new or modify existing federal laws to address confidentiality of new data sources and types
	• Determine whether existing liability protection for PSAPs and service providers will continue to apply in the NG9-1-1 environment
	• Approve and extend, if necessary, liability protection and parity for PSAPs and new service providers
	• Ensure continued access to 9-1-1 system on current and future devices and services in which users reasonably expect to access to 9-1-1
	• Consider regulations ensuring that deaf and hard of hearing individuals can be efficiently connected to 9-1-1 when such NG9-1-1 capability is available.
	• Provide clarity and update on federal regulatory boundaries and authority.
	• Educate the public and policymakers on the availability of NG9-1-1 capabilities.
	• Consider additional legislative, regulatory, of financial incentive to ensure that states and authorities provide individuals with disabilities access to the 9-1-1 system using the current and future technology and devices that are used for daily communications.

• State Government	
• Funding	• Ensure NG9-1-1 upgrades are considered a fiscal priority for states and local jurisdictions and the Federal Government
	• Consider legislation that identifies a state agency or other effective state-level mechanism (where one does not exist already) to be responsible for statewide 9-1-1 planning and granting it appropriate authority and power.
	• Encourage statewide 9-1-1 plans to justify investments for upgrading critical emergency communications infrastructure for NG9-1-1.
	• Transform current 9-1-1 funding mechanisms to resolve— • Diminishing revenue base Disparities with cost recovery • Funding allocation models for shared resources
	• Consider enacting legislation that imposes the 9-1-1 surcharge in a technologically neutral manner to accommodate current and future devices and services (e.g., text messaging, prepaid wireless).
	• Identify opportunities to offset the cost of NG9-1-1 by sharing infrastructure, resources, and services with, or simply interconnecting with, other public safety, non-public safety government, or private sector entities.
	• Consider enacting legislation that requires leveraging economies of scale to ensure efficient use of 9-1-1 revenues, and conducting annual audits on the use of the 9-1-1 funds.
	• Leverage alternative funding models (e.g., surcharge on access infrastructure provider [AIP]; user (incident) fee; Universal Statewide Communications Surcharge; General Fund tax revenue) that best suit a state's needs.
	• Consider legislation that allows 9-1-1 fees to be used to pay for the state's NG9-1-1 portion and use of IP-based emergency service internetwork.
	• Revisit how cost recovery is allocated to ensure fairness across all technologies and services, and determine whether cost recovery can or should be provided.
	• Ensure 9-1-1 funds are preserved for 9-1-1 and emergency communication systems
	• Require 9-1-1 funds to be used exclusively for the provision of 9-1-1 emergency communications services (e.g., 9-1-1, E9-1-1, NG9-1-1, and related or future upgrades).
	• Identify the appropriate uses of 9-1-1 funds and then monitor 9-1-1 Authorities to ensure 9-1-1 funds are used for costs solely attributable to 9-1-1.

• Operations	• Prepare and train call takers to handle increased quantity and quality of information available with a NG9-1-1 call
	• Where applicable, update or develop state-level standard training requirements for call takers (e.g., 40 hours of training to maintain certification).Consider conducting a nationwide annual survey to obtain input from call takers and their immediate supervisors on the effectiveness of training, SOPs, and system operations to ensure the quality of service for handling 9-1-1 calls is maintained.
• Standards and Technology	• Complete and accept NG9-1-1 open standards and understand future technology trends to encourage system interoperability and emergency data sharing
	• Require 9-1-1 Authorities to procure equipment that has been certified to meet the open standards requirements.
	• Determine responsible entity and mechanisms for location determination and acquisition
	• Determine the responsibility for location validation and management of national, regional, and/or local databases necessary for NG9-1-1.
	• Determine who within the service provider community will be responsible for providing accurate location information for NG9-1-1 calls within state.
• Governance and Policy	• Clarify jurisdictional frameworks, responsibilities, and coordination required at all levels of government to enable NG9-1-1
	• � Convene appropriate state level officials (e.g., state 9-1-1 Administrator, state homeland security/emergency management director, and state CIO) to plan for establishment of NG9-1-1 and its underlying emergency services internetwork.
	• Designate a central coordinating body and/or mechanism for NG9-1-1 implementation for each state (or region) where one does not already exist.
	• Develop a statewide NG9-1-1 implementation plan based on effective stakeholder involvement, including identifying responsible parties, goals, and milestones.
	• Consider modifying current legislation to assign state 9-1-1 Authorities with the responsibility and authority to transition to and manage the NG9-1-1 system.
	• Consider modifying current legislation to assign public utility commissions the authority and responsibility for determining certification requirements for telecommunications services and service providers that may access NG9-1-1 as appropriate at the state level.
	• Coordinate business rules and data rights management at all levels of stakeholders.
	• Update regulations, legislation, and other policies to— • Reflect communications capabilities in an NG9-1-1 system • Protect the confidentiality of new types of PII • Extend liability protection and parity for PSAPs and new services
	• Review existing laws and determine whether the existing law addresses the confidentiality of new data sources and types (e.g., medical) that could be considered PII
	• Develop new or modify existing state laws to address confidentiality of new data sources and types

	• Consider updating laws that prohibit the transmission of non-human initiated calls to a PSAP absent human intervention to accommodate sensor alerts from verified sources
	• Update state laws that prohibit 9-1-1 service from being provided by entities other than tariffed LECs/Competitive Local Exchange Carrier (CLEC)
	• Review and update laws to ensure existing statutes do not prohibit use of "virtual PSAPs."
	• Update state laws that may prohibit the use of customer/device location information for routing of other calls beyond those covered under traditional 9-1-1 (e.g., include appropriate N-1-1 entities and 800-type emergency provisioning, such as poison control centers and suicide hotlines).
	• Determine whether existing liability protection for PSAPs and service providers will continue to apply in the NG9-1-1 environment.
	• Approve and extend, if necessary, liability protection and parity for PSAPs and new service providers.
	• Ensure continued access to 9-1-1 system on current and future devices and services in which users reasonably expect to access to 9-1-1
	• Educate the public and policymakers on the availability of NG9-1-1 capabilities.
• 9-1-1 Authorities and PSAP Administrators	
• Funding	• Ensure NG9-1-1 upgrades are considered a fiscal priority for states and local jurisdictions and the Federal Government
	• Coordinate the development of statewide 9-1-1 plans and investment requests for upgrading critical emergency communications infrastructure for NG9-1-1.
	• Educate state and federal decision-makers on the importance of NG9-1-1 funding.
	• Transform current 9-1-1 funding mechanisms to resolve— • Diminishing revenue base • Disparities with cost recovery • Funding allocation models for shared resources
	• Work with state government to revisit how cost recovery is allocated to ensure fairness across all technologies and services, and determine whether cost recovery can or should be provided.
	• Ensure 9-1-1 funds are preserved for 9-1-1 and emergency communication systems
	• Strengthen checks and balances to ensure funds are used only for the provision of 9-1-1 services, including shared services (e.g., 9-1-1 portion of emergency services internetwork).
• Operations	• Prepare and train call takers to handle increased quantity and quality of information available with a NG9-1-1 call
	• Adopt training standards for processing NG9-1-1 call types, plans, implement training programs, and establish personnel qualifications.
	• Consider conducting a survey within the jurisdiction to obtain input from call takers on the effectiveness of training, SOPs, and system operations to ensure the quality of service for handling 9-1-1 calls is maintained.

	• Participate in and provide feedback on any state-level call-taker training.
	• Ensure regular training and exercises are completed at the PSAP.
	• Prepare 9-1-1 Authorities to handle NG9-1-1 system administration, including configuration management, database management, quality assurance, and SOPs
	• Develop and execute MOUs so that 9-1-1 Authorities can work together (across state and jurisdictional limits) to determine processes for call congestion, load sharing, backup conditions, and other automatic routing conditions.
	• Where needed, redefine roles and responsibilities for NG9-1-1 system administration.
	• Test the NG9-1-1 system to ensure that the level of quality of service is maintained based on terms and conditions agreed upon with the service providers.
	• Test the NG9-1-1 client application (e.g., HMI) to ensure that the level of quality of service is maintained based on terms and conditions agreed upon with the service providers.
	• Assess the amount of resources, both physical and human, necessary to provide NG9-1-1 coverage/service to an area based on population, number of calls, and other factors.
	• Prepare 9-1-1 Authorities and PSAP Administrators to handle contingency planning and use of "virtual PSAPs"
	• Develop requirements for virtual PSAPs within their jurisdiction.
	• Develop requirements and options for NG9-1-1 contingency planning (e.g., including backup, overflow, redundancy issues) for their jurisdiction.
	• Participate in operational and technical working groups to help define model requirements for virtual PSAPs and contingency requirements.
	• Hold regular combined training exercises to test the contingency plans and work to develop continuity plans of the desired level of detail.
• Government and Policy	• Clarify jurisdictional frameworks, responsibilities, and coordination required at all levels of government to enable NG9-1-1
	• Participate in the development of a statewide NG9-1-1 implementation plan, including identifying responsible parties, goals, and milestones.
	• Work with federal, state, and local level entities to coordinate business rules and data rights management within the appropriate jurisdictional boundaries.
	• Ensure continued access to 9-1-1 system on current and future devices and services in which users reasonably expect to access to 9-1-1
	• Educate the public and policymakers on the availability of NG9-1-1 capabilities.

• Public Safety Communications Organization and Associations	
• Operations	• Prepare and train call takers to handle increased quantity and quality of information available with a NG9-1-1 call
	• Develop guidelines for personnel skills/qualifications and effective training programs, including model training requirements for processing NG9-1-1 call (e.g., call handling, call treatment, and records management protocols).
	• Develop models for sharing data and managing information among PSAPs, public safety responders, and other authorized stakeholders.
	• Develop standardized common terminology that can be used by call takers similar to ongoing national standardization efforts involving various responder entities and others.
	• Consider conducting an annual survey to obtain input from call takers and their immediate supervisors on the effectiveness of training, SOPs, and system operations to ensure the quality of service is maintained.
	• Prepare 9-1-1 Authorities to handle NG9-1-1 system administration, including configuration management, database management, quality assurance, and SOPs
	• Develop a generic Memorandum of Understanding (MOU) and other related templates so that 9-1-1 Authorities can work together (across state and jurisdictional limits) to determine processes for call overflows, backup conditions, and other automatic routing conditions.
	• Develop and gather implementation, operations, and maintenance best practices and standards, and share them among 9-1-1 Authorities.
	• Develop SOPs, protocols, and definitions for system configuration and management of different call types (e.g., text message).
	• Prepare 9-1-1 Authorities and PSAP Administrators to handle contingency planning and use of "virtual PSAPs"
	• Develop requirements for virtual PSAPs.
	• Develop requirements and options for NG9-1-1 contingency planning (e.g., including backup, overflow, redundancy issues).
• Standards Development Organizations	
• Standards and Technology	• Complete and accept NG9-1-1 open standards and understand future technology trends to encourage system interoperability and emergency data sharing
	• Work with the 9-1-1 community to identify critical standards, future technologies, and all relevant data generators and user communities based on an industry-accepted architectural model and detailed requirements.
	• Conduct a gap analysis of existing standards development work against the industry-accepted architectural model and identify standards that will need to be developed and/or already developed standards that are inconsistent with the model.

	• Develop a certification program to encourage and verify that NG9-1-1 services and equipment meet the open standards defined by the industry-accepted architectural model.
	• Strengthen rules and procedures to ensure that NG9-1-1 essential standards and technology development occur in an open, fair, and competitively neutral environment (recognizing the nature of technology convergence, and competitive interests involved).
	• Establish system access and security controls to protect and manage access to the NG9-1-1 system of systems
	• Develop a certification and authentication process to encourage service providers and 9-1-1 Authorities meet the security and system access requirements.
	• Identify and leverage defense and homeland security solutions, standards, and best practices to establish security and system access standards and requirements for NG9-1-1.
	• Define identity and data rights access and management standards that can ensure only specific entities or individuals may access the NG9-1-1 system and data.
	• Determine responsible entity and mechanisms for location determination and acquisition
	• Develop data standardization for the delivery of location information whether it is civic or geospatial.
	• Determine routing and prioritization protocols and business rules
	• Identify from the user community what types of data are needed and standardize the types of data, decisions, and/or policies that should be used to route and prioritize NG9-1-1 calls.
	• Develop nationally defined protocols and business rules and options (either decision and/or policy based).
• State Utility Commissions	
• Standards andTechnology	• Establish system access and security controls to protect and manage access to the NG9-1-1 system of systems
	• Coordinate and modify existing state regulations to allow service providers beyond the common carriers to access the NG9-1-1 system.
	• Identify and modify, as necessary, state contractual provisions that may constrain the transition to the NG9-1-1 environment.
	• Assist SDOs in developing a certification and authentication process to ensure service providers and 9-1-1 Authorities meet security and system access requirements.
• Responder Agencies	
• Operations	• Prepare 9-1-1 Authorities to handle NG9-1-1 system administration, including configuration management, database management, quality assurance, and SOPs
	• Determine what data have the greatest utility in enhancing response and what processes will be used to deliver the data to ensure the right information gets to the right person at the right time.

• Service and Equipment Providers	
• Standards and Technology	• Complete and accept NG9-1-1 open standards and understand future technology trends to encourage system interoperability and emergency data sharing
	• Participate in the definition and development of open standards for the NG9-1-1 System and equipment.
	• Develop NG9-1-1 equipment and update telecommunications services to meet the open standards identified by the 9-1-1 community and federal government to promote interoperability.
	• Participate in certification of equipment to ensure the equipment meets the open standard requirements set forth by the 9-1-1 community and federal government.
	• Determine responsible entity and mechanisms for location determination and acquisition
	• Develop methods or modify existing methods to obtain the necessary location information for all call types that access NG9-1-1 and provide as mandated by the federal government.
• Government and Policy	• Ensure continued access to 9-1-1 system on current and future devices and services in which users reasonably expect to access to 9-1-1
	• Educate the public and policymakers on the availability of NG9-1-1 capabilities on telecommunications devices, as well as services used (e.g., VoIP). Similar to VoIP providers informing customers of the limitations of 9-1-1 capabilities with VoIP service.
• General Public	
• Government and Policy	• Ensure continued access to 9-1-1 system on current and future devices and services in which users reasonably expects to access to 9-1-1
	• Understand and be aware of available NG9-1-1 capabilities and coverage.
	• Provide recommendations and input into the NG9-1-1 development and deployment process to account for community needs.

REFERENCES

[1] It is assumed that emergency service networks will be IP-based and shared with other emergency and government services. NG9-1-1 service must be a priority service on the shared IP networks.

[2] USDOT ITS JPO, *Next Generation 9-1-1 (NG9-1-1) System Initiative: Concept of Operations*, April 2007, available athttp://www.its.dot.gov/ng9-1-1/pdf/NG9-1-1ConOps_April07.pdf

[3] Federal Communications Commission, Report and Order and Further Notice of Proposed Rulemaking, Adopted June 12, 1996, <http://www.fcc.gov/Bureaus/Wireless/ Orders/1996/fcc96264.txt> (last accessed March 11, 2008).

[4] Federal Communications Commission, Report and Order and Further Notice of Proposed Rulemaking, Adopted May 19, 2005. <http://hraunfoss.fcc.gov/edocs_public/ attachmatch/FCC-05-116A1.pdf> (last accessed March 11, 2008).

[5] National Emergency Number Association, *9-1-1 Fast Facts,* February 4, 2008,<http://www.nena.org/pages/Content.asp?CID=144&CTID=22> (last accessed March 11, 2008).

[6] Ibid.

[7] Ibid.

[8] As a *New York Times* article observed: The 9-1-1 "system has not kept pace with the nation's rapidly changing communications habits. As it ages, it is cracking, with problems like system overload, understaffing, misrouted calls, and bug-ridden databases leading to unanswered calls and dangerous errors." Shaila Dewan, *An SOS for 911 Systems in Age of High-Tech.* (New York Times, April 6, 2007) <http://www.nytimes.com/2007/04/06/us/06phone.html> (last accessed March 11, 2008).

[9] FCC, *Revision of the Commission's Rules to Ensure Compatibility with Enhanced 9-1-1 Emergency Calling Systems, FCC Docket No. 94-102, RM-8143, Report and Order and Further Notice of Proposed Rulemaking*, June 1996, available at <http://hraunfoss.fcc.gov/edocs_public/attachmatch/FCC-96-264A1.pdf> (last accessed March 11, 2008).

[10] For more information, please consult the U.S. Department of Justice, Civil Rights Division, *Americans with Disabilities Act:Access for 9-1-1 and Telephone Emergency Services* available at: <http://www.usdoj.gov/crt/ada/911ta.pdf> (last accessedMarch 11, 2008)

[11] "Emergency Services Internetwork" is an element of the NG9-1-1 Community Model that illustrates the 9-1-1 systems, applications, and information repositories that seamlessly share emergency data to improve response." FCC NRIC VII Focus Group 1D, *Communications Issues for Emergency Communications Beyond 9-1-1, Report #1,* December 6, 2004, available at <http://www.nric.org/meetings/docs/ meeting_ 20041206/FG1D%20Final%20Report.pdf> (last accessed March 11, 2008).

[12] "Convergence" is the integration of traditional telecommunications and newer IT services.

[13] USDOT ITS JPO, *Next Generation 9-1-1 (NG9-1-1) System Initiative: Concept of Operations*, April 2007, available at http://www.its.dot.gov/ng9-1-1/pdf/NG9-1-1ConOps_April07.pdf

[14] USDOT ITS JPO, *Next Generation 9-1-1 (NG9-1-1) System Initiative: Architecture Analysis Report*, November 2007, available at http://www.its.dot.gov/ng911/ pdf/1.F2_FINAL_MED_ ArchitectureAnalysis_v1.0.pdf

[15] USDOT ITS JPO, *Next Generation 9-1-1 (NG9-1-1) System Initiative: Concept of Operations*, April 2007, available at http://www.its.dot.gov/ng911/pdf/NG911 ConOps_April07.pdf

[16] A concept by where NG9-1-1 makes the physical, geographic location of a PSAP immaterial. Through operational, technical, and networking planning, PSAPs may acquire and leverage resources outside of their normal environment. For example, a single PSAP could experience a dramatic and rapid increase in call volume due to a large scale incident occurring in their locale. Using NG9-1-1, their pre-planned PSAP partners could provide additional technologic and staff support to handle the call volume, even though they are physically located hundreds (or thousands) of miles away.

[17] USDOT ITS JPO, *Next Generation 9-1-1 (NG9-1-1) System Initiative: Concept of Operations*, April 2007, available at http://www.its.dot.gov/ng9-1-1/pdf/NG9-1-1ConOps_April07.pdf

[18] A "virtual facility" (e.g., emergency command vehicle, backup facility, or other ad hoc location) equipped and staffed to receive 9-1-1 calls that augments an existing municipal or county emergency communications center or 9-1-1 call center that directs 9-1-1 or other emergency calls to appropriate police, fire, and emergency medical services agencies and personnel.

[19] 108th Congress, Public Law 108-494, *Enhance 9-1-1 Services*, December 23, 2004, available athttp://frwebgate.access.gpo.gov/cgi bin/getdoc.cgi?dbname=108_cong_ public_laws&docid=f:publ494.108.pdf

[20] The PSAP Administrator directs the overall operation of a PSAP and is responsible for the direct supervision, training, and administration of the PSAP's staff. The PSAP Administrator may be responsible for the maintenance of PSAP call-taking equipment and supporting peripherals. The PSAP Administrator may be responsible for the PSAP's budget and staff support. *USDOT NG9-1-1 System Description and High-Level Requirements Document,* July 31, 2007. <http://www.its.dot.gov/ng911/pdf/ NG911_HI_RES_Requirements_v2_20071010.pdf>

[21] NENA maintains several standing working groups within its Operational Development Committee (ODC). Please see the associated website <http://www.nena.org/pages/ ContentList.asp?CTID=15> for more information. 22

[22] An HMI is a PSAP user interface that will allow the call taker to quickly and intuitively interpret the data and forward asappropriate.

[23] More information about the OASIS Emergency Management Technical Committee is available at: <http://www.oasisopen.org/committees/emergency/>. For more information on EDXL see: <http://xml.coverpages.org/edxl.html> 24

[24] USDOT ITS JPO, *Next Generation 9-1-1 (NG9-1-1) System Initiative: Architecture Analysis Report*, November 2007, available at: http://www.its.dot.gov/ng911/pdf/ 1.F2_FINAL_MED_ArchitectureAnalysis_v1.0.pdf 25

[25] As of February 27, 2008, the U.S. Senate had passed this bill, and the House of Representatives must still take action. For more information, see: <http://www.govtrack.us/congress/bill.xpd?bill=s110-428>

[26] FCC NRIC VII Focus Group 1B, Properties of Emergency Communications Network by 2010, September 23, 2004, available at http://www.nric.org/meetings/docs/ meeting_20040923/NRIC%20VII%20Focus%20Group%201B%20Report_ Sept.%20v10%20_1 20304.pdf.

[27] A 9-1-1 authority has jurisdiction over and/or supports a particular 9-1-1 system or PSAP. The 9-1-1authoritycould be a county/parish or city government, a special 9-1-1 or Emergency Communications district, aCouncil of Governments, an individual PSAP, a statewide agency, or other similar body.

[28] USDOT ITS JPO, *Next Generation 9-1-1 (NG9-1-1) System Initiative: Concept of Operations*, April2007,available at http://www.its.dot.gov/ng911/pdf/ NG911Con Ops_April07.pdf

[29] Duke Law and Technology Review, *Enhanced 9-1-1 Technology and Privacy Concerns: How Has the Balance Changed Since September 11?*, October 26, 2001, available at http://www.law.duke.edu/journals/dltr/articles/pdf/2001DLTR0038.pdf

[30] In addition to essential information that is necessary to route the call appropriately, call takers (at most PSAPs) may receive a wide range of additional information that is categorized as supportive or supplemental. Supportive data is call data that may support call handling and dispatch beyond essential data (e.g., ACN data). Supplemental data is call data that

may complement, but is not necessary for, call handling and dispatch or emergency response (e.g., allergies to certain medications).

[31] Call records management involves creating, logging, archiving, retrieving, and transmitting call records. A call record is the complete, end-to-end record of a call and includes the voice recording, text communications, and essential and supplemental data associated with a call, as well as information added by the call taker.

[32] An NG9-1-1 call will potentially also include considerably more call stream data than is provided in traditional E9-1-1 (which is basically a 10-digit telephone number or, in the case of some wireless 9-1-1 setups, a 20-digit number).

[33] "Greenfield" 9-1-1 refers to PSAPs that lack a call center and only provides 9-1-1 service throughremote call forwarding. See Appendix A, p. A-1 for definitions of the PSAP categories.

In: Next Generation 9-1-1
Editor: Eugene B. Zelik
ISBN: 978-1-60692-639-0

Chapter 2

NEXT GENERATION 9-1-1 FREQUENTLY ASKED QUESTIONS

RITA, U.S. Department of Transportation

Q: What is the Next Generation 9-1-1 Initiative?

A: The Next Generation (NG9-1-1) Initiative is a US Department of Transportation (USDOT) Research and Development project funded by the Intelligent Transportation System (ITS), Joint Program Office (JPO), that will define a system architecture for emergency communication that enables the transmission of voice, data or video from different types of communication devices to Public Safety Answering Points (PSAPs) and onto emergency responder networks. The NG91-1 Initiative will also develop a transition plan that considers responsibilities, costs, schedule and benefits for deploying this architecture.

Q: What's wrong with the current 9-1-1 system?

A: The infrastructure of the Nation's current 9-1-1 system has essentially not changed since the first 9-1-1 call was made in 1968. Connections to this network have been adapted to accommodate TTY and wireless calls, but the basic infrastructure is decades old and cannot handle the text, data, images and video that are increasingly common in personal communications and critical to future transportation safety and mobility advances. To date, there has been no planned method for receiving the additional available data that could affect or improve the delivery of emergency services

The NG9-1-1 Initiative will establish a foundation for emergency communications services in a wireless mobile society. It will create a framework for 9-1-1 services to advance from analog to digital technology, and from copper landlines to Internet Protocol (IP)-based internetworks. This will enable the transmission of voice, text, video and photos, and will result in a more robust, resilient 9-1-1 network.

Q: How is the US Department of Transportation involved?

A: The NG9-1-1 Initiative is co-managed by the ITS JPO and the Office of Emergency Medical Services at the National Highway Traffic Safety Administration (NHTSA). It is leveraging work from USDOT's earlier Wireless E9-1-1 Initiative, as well as myriad of other requirements and technical approaches undertaken by both public and private entities, representing the public safety community as well as the work done by industry and government.

Q: Who is involved in the NG9-1-1 Initiative?

A: USDOT is leading a project team through a contract with the firm of Booz Allen Hamilton (BAH) as the primary contractor. The NG9-1-1 Initiative team also includes the National Emergency Number Association (NENA), L. Robert Kimball & Associates as well as an academic-based consortium, being led by the Texas A&M University.

Stakeholder involvement in the project is crucial its overall success. Throughout the Initiative, input is being solicited from a large group of stakeholders. The 9-1-1 constituency is a diverse group of entities, including local, State and Federal government agencies, professional and industry associations, Standards Development Organizations, consumer advocacy organizations, research and academia, traditional, IT and VoIP telecommunication service providers and companies providing hardware and software to PSAPs. These stakeholders are being involved in all aspects of the project dealing with the creation, delivery, receipt, and management of 9-1-1 calls.

Q: What is the timeline for the NG9-1-1 initiative?

A: The current NG9-1-1 Initiative is expected to run from December 2006 through November 2008.

Q: How is the NG9-1-1 Initiative organized?

A: Both the DOT and the Booz Allen team have distinct responsibilities. The DOT began the work by completing an initial Concept of Operations document and engaging the stakeholders.

There are four (4) distinct tasks, or phases, included in the Booz Allen team contract, including:

Task 1: System Architecture,
Task 2: Preliminary Transition Planning,
Task 3: Proof of Concept Demonstration, and
Task 4: Final Transition Planning.

Each of these tasks has subtasks and is build upon one another and work on multiple tasks will proceed in parallel. Once the Booz Allen team completes the four tasks, the results of the NG9-1-1 Initiative will transition to the National 9-1-1 Office.

TASK 1: SYSTEM DEVELOPMENT

Q: What work will be accomplished during Task 1, System Architecture ?

A: The following subtasks are included in Task 1:

Revision of Concept of Operations,
High Level Requirements,
Detailed System Requirements,
Configuration Management, and
Architecture Analysis.

Q: What is the Concept of Operations document?

A: The purpose of the Concept of Operations ("ConOps") is to communicate the vision of the NG9-1-1 system and provide a high-level description of what the NG9-1-1 system will do. It serves as the foundation for the development of the NG9-1-1 requirements and will drive the design of the o verall system architecture and transition plan. The intended audience for the ConOps includes the users, stakeholders, and the operational and support elements that comprise the NG9-1-1 System. The ConOps is intended to provide a user-oriented vision of NG9-1-1 that can be understood by stakeholders with a broad range of operational and technical expertise so that they can actively engage in its development and deployment.

Q: What is meant by the "Emergency Services Internetwork"?

A: The "Emergency Services Internetwork," as defined by the Network Reliability and Interoperability Council (NRIC) VII Focus Group 1D, refers to a network of networks, or a "series of secure local, regional, and national wireline and wireless [9-1-1] networks providing modern, integrated [emergency] information capabilities to support local, regional, and national needs, or a system of systems".

Q: What is a Community Model and what is its purpose?

A: The NG9-1-1 Community Model is a graphic model depicting users, stakeholders, and the associated operational and support elements that comprise the NG91-1 System. The Community Model allows these stakeholders to visualize how the NG9-1-1 System interacts with all of these entities and how they fit into the emergency communication internetwork or network of networks.

Q: What are requirements?

A: Beginning with the vision of the ConOps document, high-level requirements are developed with input from stakeholders who will use, operate and support the NG9-1-1 architecture. The requirements describe what functions the system must perform, not how the system carries out these functions. The requirements definition process is focused on capturing and translating the NG9-1-1 mission into functional, system, performance, and data requirements.

Q: Which Standard Development Organizations (SDOs) and other stakeholder activities are being considered?

A: Key documents and activities will be considered from a variety of SDO sources including:

the Internet Engineering Task Force's (IETF),
the Emergency Context Resolution with Internet Technologies (ECRIT), and
the Alliance for Telecommunications Industry Solutions' (ATIS) Emergency Services Interconnection Forum (ESIF).

Information from other documents and resources has also been incorporated as part of the starting point for the NG9-1-1 Initiative, including:

The National Emergency Number Association's (NENA) i3 Technical Requirements and

The Federal Communication Commission's (FCC's) National Reliability and Interoperability Council (NRIC) Focus Groups 1B and 1D.

TASK 2: PRELIMINARY TRANSITION ANALYSIS

Q: What work will be accomplished during Task 2, Preliminary Transition Analysis?

A: The following subtasks are included in Task 2:

Preliminary Analysis of Cost, Value, and Risk and
Preliminary Transition Issues Report and Plan.

Q: How will the Analysis of Cost, Value and Risk be developed?

A: Completion of the Cost-Value-Risk Analysis involves developing a framework and a process to evaluate the life-cycle costs, benefits and risks for each architecture alternative, ranging from doing nothing (i.e., status quo or as-is architecture), to implementing the "to be" NG9-1-1 architecture. It includes a qualitative estimate of non-financial value that may not be accounted for in traditional cost/benefit analysis, allowing for a more rigorous comparison of alternatives.

Q: How will the Transition Plan be developed?

A: The transition plan will identify critical system deployment issues, including a variety of operational, regulatory, institutional, governance, funding, and public policy issues critical to successful transition to a next generation IP-based architecture. Working with stakeholders, the Final Transition Plan will describe strategies and models as options for overcoming challenges to deployment.

TASK 3: PROOF-OF-CONCEPT DEMONSTRATION

Q: What work will be accomplished during Task 3, Proof of Concept (POC)?

A: The following subtasks are included in Task 3:

POC Deployment Plan,
Call Taker Human Machine Interface (HMI) Design,
Data Analysis Plan/Data Acquisition System Design,
Proof-of-Concept Test Plan ,
POC Demonstration Report/Test Results, and
Final POC System Design.

Q: What are the objectives of the Proof-of-Concept?

A: The objective of the Proof-of-Concept is to test specific requirements from the NG9-1-1 Detailed System Requirements Document. Specific requirements were selected from a prioritized list developed by the NG9-1-1 team with input from a variety of 9-1-1 stakeholders. The specific objectives of the POC include:

Developing a working NG9-1-1 model that can receive text, images, and video (in addition to voice) from several different kinds of access networks and communications devices (e.g. PDAs, computers, auto telematics),

Demonstrate 9-1-1 call transfer, backup and distance access among geographically dispersed PSAPs using NG9-1-1, IP-based routing capability, and

Demonstrate use of open platforms and non-proprietary standards for developing NG9-1-1 solutions.

Q: What is the timeframe of the Proof-of-Concept?

A: The POC is expected to begin in April 2008 and will last approximately two to three months. Following completion, data gathered during the POC will be analyzed and used to revise the system design and complete the transition plan.

Q: How will the Proof-of-Concept be conducted?

A: Using all of the documents created in preceding tasks, the project will construct, implement and test selected, specific system requirements of the NG9-1-1 network; using a combination of "live" and "lab" facilities. The POC will include three lab facilities and five PSAP locations. While live data will not be used, PSAPs will provide access to varying equipment and staff to participate in the POC.

Q: Where are the three lab facilities located?

A: The three lab facilities that will be used for the POC are:
Booz Allen's Center for Network and Systems Innovation (Herndon, VA),
Texas A&M Internet2 Technical Evaluation Center (College Station, TX),
and Columbia University (New York City, NY).

Q: How were the PSAP locations selected by USDOT to be participants in the POC? Who was selected?

A: Five PSAP locations were selected from over 50 applicants, using objective criteria developed by the NG9-1-1 team. While the field included many impressive applicants, resulting in a very close competition, the NG9-1-1 Initiative is limited by funding and schedule, requiring the decision to limit participation. The USDOT sincerely acknowledges and appreciates the willingness of all applicants to go "above and beyond" in offering to participate in the POC. The selected PSAPs are:

City of Rochester – Emergency Communications Department, Rochester, NY,
King County E-911 System, Seattle, WA,
Metropolitan Emergency Services Board – Ramsey Co. Emergency Communications Center, St. Paul, MN,
State of Montana – Public Safety Services Bureau, Helena, MT, and
State of Indiana - Office of State Treasurer, Indiana Wireless 911 Board.

Q: What is included in the Final POC System Design Assessment?

A: The System Design Assessment includes an evaluation and a record of all problems, defects, and errors discovered during the proof-of-concept demonstration as defined in the Proof-of-Concept Test Plan. Additionally, the overall deployment experience will be evaluated to identify enhancements that will improve the system design (e.g., reduce operating costs, update user interface, and simplify data access).

TASK 4: FINAL TRANSITION ANALYSIS

Q: What work will be accomplished during Task 4, Final Transition Plan?

A: The following subtasks are included in Task 4:

Final Cost-Value-Risk Analysis and
The Final Transition Plan.

Q: How will the Final Transition Analysis be conducted?

A: This final version of the Transition Plan will draw upon the findings and conclusions of Tasks 1, 2, and 3 as well as input solicited from stakeholders to further identify and refine deployment issues and describe strategies to address identified issues.

Q: How can I find out more about the NG9-1-1 Initiative?

A: A complete description of the NG9-1-1 Initiative as well as completed documents can be found at http://www.its.dot.gov/ng911.

Updated June 12, 2008 10:44 AM

INDEX

D

L

M

N

T